GENERATING PRESSURE

The Campaign against Nuclear Power at Druridge Bay

Bridget Gubbins
for the Druridge Bay Campaign

Illustrations Jane Gifford

Druridge Bay View. *Photo Hugh Stephenson*

AUTHOR'S ACKNOWLEDGMENTS

This story is a result of the combined efforts of many people. I would like to give special thanks to the following:

- the Steering Committee of the Druridge Bay Campaign for asking me to write this story, and giving me the time as part of my work to do it.

- Fiona Hall, DBC Vice-Chair and writer, who has closely collaborated with me from its conception to completion. Also to Mike Kirkup who helped with editing.

- Jonathon Porritt for writing the foreword, and for the inspiration he has given us over the years.

- Northumberland County Library Service, who have supported this work from the beginning.

- Earthright Publications for their belief in the value of our story.

- local authorities in Northumberland, Tyneside and County Durham, without whose financial support the Druridge Bay Campaign could not sustain its activities on behalf of the people of the north-east.

- politicians of all colours who have united behind our struggle, and who I confidently expect will do so in the future.

- reporters and editors of local newspapers, radio and TV, who reported on our activities willingly and impartially.

- the many people who answered my questions about their personal involvement, and those who supplied me with photos.

- the many people not personally mentioned in this book whom I know have contributed directly to the campaign.

- the people of the north-east, who have never kept secret their anger about the defiling of Druridge Bay, and whose voting has reflected this.

For more information about opposition to nuclear power at Druridge Bay please contact the Druridge Bay Campaign, Tower Buildings, Oldgate, Morpeth, Northumberland NE61 1PY; tel: 0670-513513.

Cover photo: *Peter Berry*
Text © Bridget Gubbins 1991
Drawings © Jane Gifford 1991
Photos: © the photographers
Typeset by Roger Booth Associates, 10 The Bigg Market, Newcastle upon Tyne NE1 IUW
Printed on recycled paper by Tyneside Free Press, 5 Charlotte Square, Newcastle upon Tyne NE1 4XF
Published by Earthright Publications, 8 Ivy Avenue, Ryton, Tyne and Wear NE40 3PU

British Library Cataloguing-in-Publication Data
Gubbins, Bridget
    Generating pressure : the campaign against nuclear power
    at Druridge Bay.
    I. Title
    333.7924

ISBN 0 907367 08 9

CONTENTS

Foreword Jonathon Porritt	v
Chronology	vi
Map	viii
INTRODUCTION	1
1. NUCLEAR POWER AT DRURIDGE? 1978 – 1981 New pressure group, the Druridge Bay Association	3
2. FRIENDS OF THE EARTH MID NORTHUMBERLAND 1983 Reviving the opposition	24
3. DRURIDGE BAY CAMPAIGN – A FEDERATION 1984	42
4. THE CEGB IS DETERMINED 1984	48
5. RADIATION AND CHERNOBYL 1984 – 1986	60
6. POLITICAL DEVELOPMENT 1985 – 1987	70
7. THE ARGUMENTS 1983 – 1989 Environment, Safety, Better Alternatives, Jobs, Nuclear Waste	83
8. WHAT PEOPLE HAVE DONE 1983 – 1989	99
9. PRESSURE BUILDING UP 1988 – 1989	125
POSTSCRIPT	149

To the children of the North-East, and those of future generations.

FOREWORD   Jonathon Porritt

The very idea of a nuclear reactor at Druridge Bay seems to me to be sacrilege. Yet this dramatic stretch of Britain's shoreline, with its great sweep of sea and sky, has been under such a threat for the last 12 years.

With the privatisation of the electricity supply industry, and the government's own acknowledgment that nuclear power is hopelessly uneconomic, that threat has been temporarily lifted. But not completely removed. The Pressurised Water Reactor (PWR) at Sizewell B should be completed by the end of 1993, at a final cost of more than £2 billion. The go-ahead has been given for another PWR at Hinkley Point in Somerset, but even this now looks in doubt. A review of nuclear policy (promised for 1994) will determine its fate, and that of all the other sites threatened with a PWR. Including Druridge Bay.

Having watched the campaign at Druridge Bay from afar, occasionally taking part in some of its many activities, I read this book with a renewed feeling of astonishment at the dedication and commitment of those most directly involved in the campaign. It is an extraordinary thought that groups of similarly far-sighted and concerned people are battling it out in their own communities the length and breadth of the country, pockets of indomitable resistance to the juggernaut of mindless industrial progress.

Primarily, it is a tale about the nuclear industry. There is much to be learned from it about the issues themselves, about the arguments for and against nuclear power, and for and against a nuclear reactor at Druridge Bay.

But there's also a lot to be learned about the nature of citizen protest and the dynamics of mobilising individuals and whole communities in the art of exerting democratic pressure.

The mass media's shorthand about such environmental campaigns gives a very unrepresentative picture. It all sounds rather exciting, glamorous even, when conveyed in thirty-second sound bites. In reality, such campaigning is a long hard slog, 99 per cent of which goes on behind the scenes and beyond the reach, let alone the understanding, of the media.

The authenticity of that experience is writ large on every page of this book, with the personal voices of those most directly involved in the campaign providing a compelling testimony of what it is to become an "environmental activist". As the author says, this was not a role any of them actively sought, but one forced upon them by the blundering insensitivity of the erstwhile CEGB.

It is therefore a story of genuinely collective courage and commitment. But no collective can work without the inspiration of one or more individuals. Bridget Gubbins, the author, has too modestly hidden her own light under a bushel, but it should shine through brightly despite such self-effacement.

I have always been a sucker for "David and Goliath" stories, but an account such as this really does fill me full of hope and confidence for the future.

• • •

CHRONOLOGY

1973 **December**, South of Scotland Electricity Board (SSEB) applied to build up to 8 nuclear power stations at Torness.
1974 **June**, Public Inquiry commenced at Torness
November, Inspector's report, consent advised.
1975 **November**, Scottish Campaign to Resist the Atomic Menace (SCRAM) formed.
1978 Protests, marches, actions at Torness.
1978 **March**, Cheviots announced as possible site for high level nuclear waste dumping.
1978 **Dec 22**, ANNOUNCEMENT THAT DRURIDGE BAY COULD BE SITE FOR NUCLEAR POWER STATION
1979 **March**, Druridge Bay Association (DBA) formed.
July, DBA rally and pop concert at Druridge Bay.
August, test drilling at Druridge Bay.
December, government announced one new nuclear power station (pressurised water reactor – PWR) per year for ten years.
1980 **February**, Northumberland County Council opposed nuclear power at Druridge.
February, Cheviot Defence Action Group formed.
October, Cheviots Inquiry opened in Newcastle, closed December.
December, Central Electricity Generating Board (CEGB) declares Druridge test drilling results show Druridge is suitable for nuclear power.
1981 **May**, Northumberland County Council elections give Labour majority. Jack Thompson becomes leader.
December, high level waste storage in Cheviots postponed by government.
1982 **August**, Druridge on short list of 5 sites for PWR.
October, Northumberland County Council sets aside £100 000 to present case at Sizewell Inquiry.
1983 **January**, Friends of the Earth Mid-Northumberland founded to revive opposition to nuclear Druridge.
February, Sizewell Inquiry started. (Finished March 1985.)
March, founding of the Cairn by Friends of the Earth.
September, Gathering at Druridge Bay in storm.
November, miners' rally with Arthur Scargill in Ashington.
1984 **January**, Druridge Bay Campaign (DBC) formed as federation of opposing groups.
March, Miners' strike begins. (Ends March 1985.)
May, CEGB moves in equipment for detailed test drilling.
August, Friends of the Earth try to buy Druridge site.
December, CEGB buy Druridge land, detailed test drilling shows site is suitable for nuclear power.
1985 **March**, Sizewell Inquiry ends after 26 months.
March, miners' strike ends.

	May, Radiation Monitoring group set up by DBC.
	July, CEGB buy houses at Druridge Bay.
1986	**April**, DBC sets up office and employs first worker.
	April/May, Chernobyl accident.
	September, first Druridge Bay Campaign Fair at Druridge.
1987	**Jan**, Sizewell Inquiry report recommends go-ahead.
	Feb, DBC lobby to House of Commons on Sizewell.
	October, JOBS FORUM Gateshead Civic Centre.
1988	**June**, *Tide Lines* poetry anthology launched.
	November, DBC lobby of House of Commons on privatisation of electricity.
	December, 10 years of Druridge pressure group activity commemorated.
1989	**April**, DBC lobby of House of Lords on privatisation of electricity.
	July, Cecil Parkinson removes Magnox nuclear power stations from privatisation.
	July, launch of DBC's Greenhouse Effect argument with Jonathon Porritt.

9 NOVEMBER, DRURIDGE REPRIEVE. John Wakeham announces in House of Commons that nuclear power stations will not be privatised after all, and that NO MORE NUCLEAR POWER STATIONS WILL BE BUILT UNTIL AFTER REVIEW IN 1994.

Druridge Bay View. *Photo Sirkka-Liisa Kontinnen*

NORTH EAST ENGLAND AND DRURIDGE BAY
showing site for proposed nuclear power stations

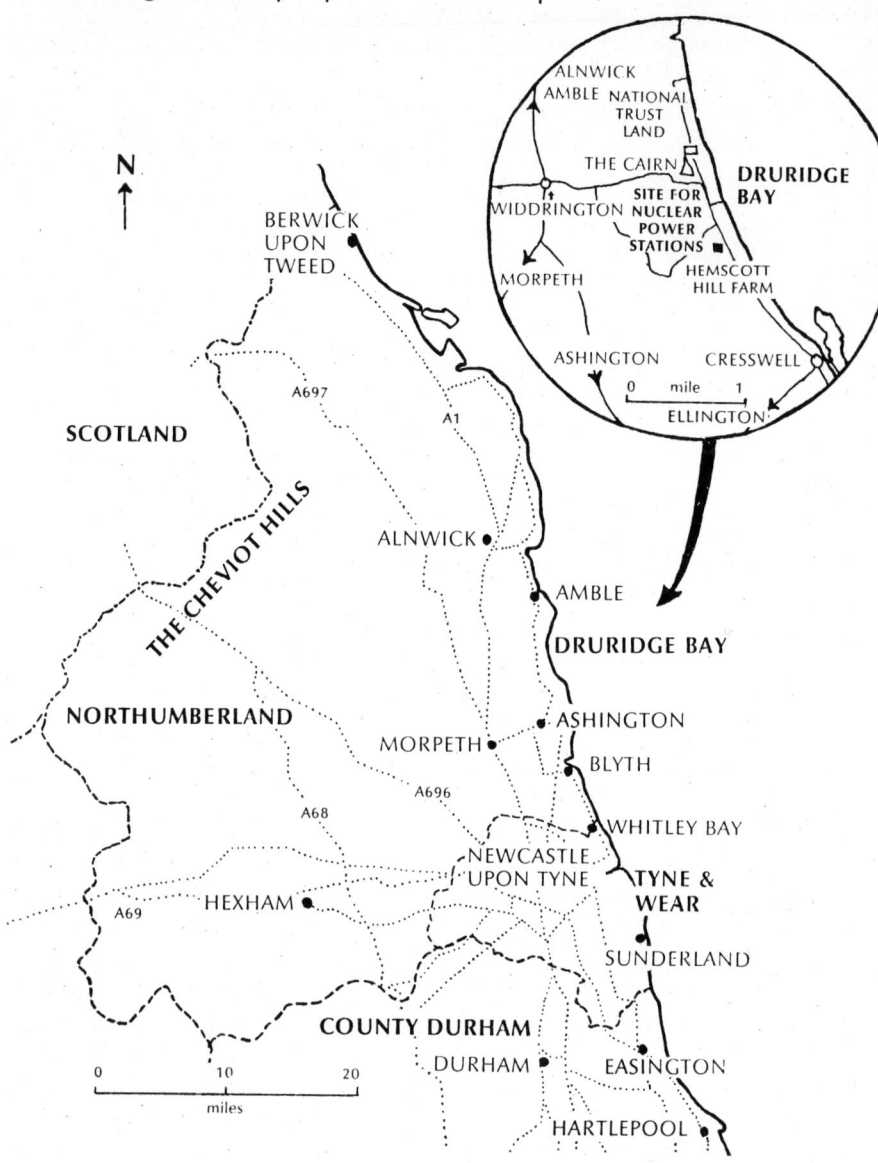

INTRODUCTION

In December 1978, the Central Electricity Generating Board (CEGB) announced to a startled public that it was interested in building a nuclear power station at Druridge Bay.

Druridge Bay is a six mile crescent of golden sands, on the windswept Northumbrian coastline, about 20 miles north of Newcastle upon Tyne. It is the most southerly of the series of great Northumbrian beaches north of Newcastle untouched in any significant way by industrial development; the closest beach for Tynesiders who seek solitude or a natural environment.

Its great open space and often chilly weather mean that it is seldom crowded. Even in the infrequent warm spells, the bay is so extensive that large numbers of people are absorbed without crowding. The glittering expanse of the North Sea, the miles of golden sands and the brilliant skyscapes are loved by thousands of people.

Evidence of coal mining at Druridge Bay which took place in the past at villages like Broomhill, Hauxley and Chevington has long since disappeared. Most of the untidy old buildings and mine workings have been removed by opencast sites at East Chevington, Hauxley, Radar and Coldrife. The land has been restored to farming, and some of it has been developed into nature reserves by Northumberland Wildlife Trust and Northumberland County Council. East Chevington opencast site will be restored by the mid 1990s, and some of this land will be used to create another bird reserve. Castle Morpeth Borough Council and the National Trust also manage large areas of duneland for public use and as nature reserves.

There is still coal suitable for opencasting in the southern section of the bay's hinterland. Local structure plans and county council policy are totally opposed to further commercial development including opencast mining at Druridge Bay. At least a million tons of coal remains under the sea, which could be deep mined. These reserves are an extension of Ellington Colliery where miners are working at present.

The tall chimneys of Blyth coal-fired power station and Alcan, clearly to be seen on the southern horizon, remind us that much of this coal goes into local industry.

The CEGB decided that Druridge Bay might be a suitable site for nuclear power stations, in spite of its beauty and the importance of coal as a source of employment. In fact they planned not one but two, with the possibility of a third. A nuclear complex was the dream of people remote from the North East, who were determined to use every method legally at their disposal to see that it came true.

Since 1979, prevailing attitudes to nuclear power have radically altered. On 9 November 1989, the same day as the Berlin Wall came down, John Wakeham, Secretary of State for Energy, declared in the House of Commons that nuclear power was not to be included in the privatisation of electricity. No more government money would be invested in new nuclear power stations until after a review in 1994.

Druridge Bay is safe – for the time being. But Nuclear Electric, the new state owned company, is determined to bring in new plans for economic nuclear power. By 1994, further proposals will emerge, and Druridge Bay will again be in danger.

This is the story of local people and their fight to save Druridge Bay, their activities and their reasoning. It tells how, from nothing, they built up a formidable organisation, generating pressure to save their environment. They have a temporary respite, time to re-assess tactics. But the threat to the Bay has not gone away. The fight will continue.

The sign at Druridge Bay, near the Cairn, erected September 1989

1. NUCLEAR POWER AT DRURIDGE?
1978 – 1981
New pressure group, the Druridge Bay Association

It was Friday, 22 December, 1978, three days before Christmas. Mothers were bringing children home from the last day of school, bustling through the dark, damp streets, getting in the groceries for the evening meal, and picking up a newspaper on the way. Once home, with children safely settled in front of the television, mothers in houses dotted across the North East sat for a few precious moments for a cup of tea and a glance at the local paper.

On the front page of the *Morpeth Herald* was a pretty design of holly surrounding the words "A Merry Christmas and a Prosperous New Year to all our Readers." Next to that was a headline, "DRURIDGE MAY BE SITE FOR A-POWER STATION". What was that? A power station? At Druridge Bay? Surely not!

"The Central Electricity Generating Board are looking at beautiful Druridge Bay as a possible site for an atomic-powered power station. They are understood to be considering test bores in the area to see whether or not it will be suitable."

An **atomic** power station? A nuclear power station? At Druridge Bay? They can't! They wouldn't! Its impossible!

Cups of tea forgotten, shock and fear penetrated one house and another. Fathers driving home heard it on the evening news. Families saw it on the telly. Councillors in Northumberland were stopped in their tracks. Meanwhile, the children clamoured for their tea, last minute Christmas cards were to be written and shopping had to be done. Kitchens had to be cleaned ready for visitors on Monday, presents had to be wrapped. Surely it couldn't be true. Nuclear power at Druridge Bay! On the beach, where people take their children! Who is proposing it? How dare they?

• • •

That is how the news was broken to the people of the North East, just before Christmas, timed so as to break the impact of one of the most dramatic announcements in the history of the area.

It seemed surprising, startling, outrageous to me. Yet, upon reflection, and with hindsight, that reaction seems foolish. Already the nuclear industry was spreading its tentacles across the North of Britain. Sellafield was undergoing a name change from Windscale, as if that would change people's perception of its persistent and insidious activities. A nuclear power station was going to be built in Torness, only 80 miles to the north of Druridge Bay. The Cheviots had been mentioned in 1978 as a potential site for the dumping of high level radioactive waste. Why should Northumberland's beaches be excluded from the list of threatened areas? The nuclear industry had seemingly no hesitation in grasping what it wanted.

Cumbria, land of lakes and mountains, loved by poets, painters, climbers, carried Sellafield's emissions in its winds. Cheviot's windswept grasslands were proposed as a radioactive rubbish dump. A clifftop site amid the greenery of agricultural land in the Scottish borders was soon to be used to build a series of nuclear power stations. Why not Druridge Bay indeed? It fitted the pattern. It was lonely. It was in the North, where votes are not so important.

But it was quite unbearable. For me, as doubtless for thousands of other local people, it seemed like the last straw.

Part of me couldn't accept it as reality, aching as I already was about the Cheviots. I tried to drive the thought away from my mind. Another part of me was dulled with fear of the hopelessness of opposing "them", those invisible people from somewhere else, with power to inflict these monstrosities upon us. And the foolish, wishful-thinking part of me dwelt on the hope that the site might prove not be suitable. After all, they were only talking about test boring.

Fortunately, in those early days of reaction to the development of nuclear power in the north of Britain, others were more sturdy, more alert, and more prepared.

The application by the South of Scotland Electricity Board in 1974 to build nuclear power stations at Torness had led to a 7-day Public Inquiry at Dunbar. Friends of the Earth were among those opposing. The Inspector recommended consent late the same year. Worried people in Lothian waited for the worst. Nothing happened for four long years. Then, in 1978, tenant farmers left their land on the Torness site, and demonstrators moved in. Bulldozers demolished their dwellings, there were many arrests, and the perimeter fence was constructed. Despite intense opposition, Torness went ahead. The opposition in those days was breaking new ground. There was no national movement against nuclear power in 1978.

Next came the Cheviot threat. In March 1978 the United Kingdom Atomic Energy Authority suggested that the Cheviot granite could be a suitable site for burying high level radioactive waste. This caused great consternation throughout Northumberland. At Torness, opposition had been prominently from politically aware left-wing anti-nuclear groups and Friends of the Earth. This time, it was country people, supporters of conservation groups, farmers and landowners, who were outraged. These people were not the kind to take up stands against the government's position on nuclear power, but they were threatened by it just the same.

Reacting to the Cheviot announcement, in November 1978 the Northumberland and Newcastle Society organised a symposium at Newcastle University, inviting expert speakers from across the world to discuss Nuclear Waste and the Environment.

Thus the rumblings of the developing nuclear industry were being felt in the North East by Christmas 1978. What would be the reaction of the general public to the proposal to build nuclear power stations at Druridge Bay? The Central Electricity Generating Board (CEGB) who had put forward the proposal for test drilling must have been watching closely, as were local politicians and concerned individuals like myself and my family and friends.

Public opinion took time to reveal itself. How does one measure it? How can one assess the undercurrent of concern, uncertainty, worry and fear within a community, which is not sure how to express itself? No doubt the CEGB carefully checked local newspapers in 1979 for people's reactions. A continuous series of articles appeared after the first announcement, not all of which were front page stories.

Castle Morpeth Borough Councillors from the local planning authority were quick to declare their opposition. In January 1979, although the Chief Executive Maurice Cole felt it was too early to make a decision, the councillors were outspoken. The Mayor, Cllr Tom Brown, said, "I for one say 'No' to such a development at Druridge Bay." Seconding him, Cllr Mrs M Brown said the local people were horrified when they heard about the proposal. Cllr D Adams, Deputy Mayor, said, "Let us show we intend to fight this." Cllr Devereux said they must fight it on every ground.[1]

Northumberland County Council was more equivocal. At that time the county council was under the control of the Northumberland Voters' Association, a Conservative/Independent alliance. "This suggestion will be looked at coolly and carefully," said the vice-chairman Cllr Lt-Colonel Robert Barnett in January. The planning committee chairman Cllr Colonel Jim Smail said that he was adamant that full public consultation would occur before any recommendation was made. The stoutly proclaimed opposition by Castle Morpeth Borough Councillors was noticeably lacking.[2]

In February, Alan Thompson, Liberal Prospective Parliamentary Candidate for Morpeth, declared his opposition, and called for a referendum from the people of Northumberland.[3]

In March, Morpeth Labour Party sent a resolution to the Regional Labour Party for debate. It said, "This conference opposes the attempt by the CEGB to establish a Nuclear Power Station at Druridge Bay. We feel that apart from environmental grounds, it would not, in the long term, benefit the employment situation which will prevail in the surrounding area of Northumberland."[4]

While the official organisations were making their views known, other unofficial moves were being made. In the same way as an alliance of the opposition was formed at Torness, and later for the Cheviots, Druridge Bay produced its own group of defenders. These new pressure groups sprang up from among concerned citizens. They were informal clusters of individuals, forming a new layer in local democracy. Not elected but self-appointed, their role was to mobilise public opinion and prod elected representatives into action.

These people didn't waste much time. They couldn't wait until councils had deliberated, or public inquiries had made recommendations. They were on the spot, ready to fight. At Druridge Bay they started a protest movement which has lasted for a decade, and which is likely to go on for many years yet.

At this stage, my husband and I were among the thousands of people in the area who were dumbstruck with concern for Druridge Bay. We had three small children, Patrick aged 8, Daniel aged 5 and Jeannie aged 1. We were consequently very busy. We were not among the early public protestors about Druridge Bay, although our concern was deeply felt. As the early campaigners carried on the initiating stages of the battle to save Druridge Bay, we carried on our daily lives, anxiously scanning the papers for odds and ends of information.

Who were the first protestors?

Sue Jordan lived in Widdrington, a mile from Druridge Bay. She worked in Newcastle City Library, as a librarian.

"I was at work indexing a newspaper, just before Christmas in 1978. Suddenly there, on the the front page of the *Journal* was Druridge Bay. A nuclear power station may be built there. I nearly dropped off my seat. My husband and I had only just moved to Widdrington, and we thought that we'd found our dream house, thought we were settled there for life. We had glorious views down to the beach. Now what was going to be in my front garden – a nuclear power station! My first reaction was to telephone my husband and weep."

Harry Allport lived at Lynemouth, a mile south of Druridge Bay. He was at *The Plough Inn* at Ellington, when he ran into Addison Brown who lived at Cresswell. Harry said, "He shouted me over and told me he'd had the electricity board down at his house, asking him and other people in the village if they would mind if they built a nuclear power station down there. I said, 'You're kidding aren't you?' He wasn't. I couldn't believe it. All these old fellows were sitting there playing dominoes, while he told me there might be a nuclear power station at Druridge Bay. I said, 'Is anybody going to do anything about it?' He said that the villagers were going to get themselves together to play hell about it."

Alan Brown lived at Widdrington. He was an engineer, working in Newcastle. He said, "When it was announced in *The Journal*, just before Christmas, I was outraged. It was Druridge Bay they were talking about. It would devastate this area. It would be horrendous. They couldn't do it."

Sue and John Jordan were having a Christmas party the night after the announcement was made. Don and Jane Kent, prominent members of Friends of the Earth on Tyneside were long standing family friends of the Jordans, and were at the party.

Alan Brown was there. He said, "I was getting on my high horse about the nuclear proposal at the party. I said, 'Somebody should do something!' Don said, 'It's up to you to do something.' I asked him what I could do. He said, 'The only way you will be taken notice of is to be spokesman or chairman of something. I suggest you have a meeting locally and form an association.' I thought it was a good idea."

Sue Jordan, Alan Brown and others quickly organised a meeting in the *Widdrington Inn*. They put leaflets and notices up in Widdrington, Cresswell and part of Ellington.

Pat Tudor lived at Cresswell, and came to the meeting. At the time, Pat was a teacher at Ashington High School. He was to become one of the new pressure group's most active members.

Pat Tudor said, "The idea of the meeting was partly to share information, but nobody had any authoritative knowledge. Don Kent from Friends of the Earth came. We did recognise that as local residents, at any planning inquiry we would have a particular role. Even though we knew we would need support from a wider area, we started at the heart of it, with the people who lived near Druridge Bay, and worked outwards. I took on the job of helping to frame a constitution for the group, as I had experience of this kind of thing working with voluntary groups."

A Steering Committee was set up at the meeting at the *Widdrington Inn*, which began to meet weekly. By March 1979, they were ready to announce to the local papers that they would present their constitution at a public meeting in Ashington on 9 April.

The new organisation was to be called the Druridge Bay Association, and was to be non-party in politics and non sectarian in religion. The constitution was adopted at this meeting, and made the aims of the group clear. It was to be a residents' or community association, and not an anti-nuclear group. Its object was as follows: "To promote the benefit of the inhabitants of the area bordered by the River Coquet to the North, the River Wansbeck to the South and the A1 road to the West." It aimed to unite people "in a common effort to advance education and to provide facilities in the interests of social welfare for recreation and leisure time occupation with the object of maintaining and improving the conditions of life for the said inhabitants."

Because of the particular way in which the Druridge Bay Association set itself up, it was able to qualify for charitable status. Although its consequent actions were mainly those of a pressure group which had the object of defending Druridge Bay from nuclear power stations, its stated aims were those of a residents' association.

Early on in its development, the committee was quite clear that the style of the work the Druridge Bay Association (DBA) undertook was to be unlike that of anti-nuclear campaigners.

The DBA wanted to present the image of ordinary grassroots people protesting to protect their area. The reason behind this was totally practical, as the members felt their campaign had to impress the public as being reasonable, or even conventional. On the other hand, the DBA was indebted to the anti-nuclear Friends of the Earth, who were able to provide them with a good deal of information. The DBA however wanted to carve out its own identity.

The decision of whether to oppose nuclear power at Druridge Bay by being anti-nuclear or not was to be taken several times in the following decade, after the forming of the Druridge Bay Association. It was always a subject of debate.

Elsie and Alf Townley lived at Blakemoor Farm, on Druridge Bay, adjoining one of the CEGB's two selected sites. Elsie was a teacher, and Alf a farmer. Both were life-long Labour Party activists, and committedly anti-nuclear. Elsie said of the DBA of which they were members, "The DBA bent over backwards to be a residents' association. They regarded groups like Friends of the Earth as the lunatic fringe."

Alan Brown put it like this. "We wanted to be different from Friends of the Earth. We wanted our concern to concentrate on Druridge Bay. We felt we had a winning line there. There was a fringe of Friends of the Earth which is classified as 'sandals and beards', and people didn't take them seriously in those days. The DBA committee were mainly business or professional people. We wanted a different approach."

Sue Jordon said, "The one way for us to retain credibility was to show people we were ordinary people, a mixture of housewives, residents, professionals; that we weren't professional protesters. By contrast, Don's Friends of the Earth would turn up at a meeting with long hair, badges, and scarves down to their feet. They really cared, and I'm sure they are still protesting about the issues. But get some journalists there taking a photograph of them, and that's your campaign smashed. We would be labelled. By now, Friends of the Earth has established a very good reputation. But back then it was different."

I asked Don Kent recently to respond to these comments. He said, "I saw the furore about Druridge Bay as part of the broader anti-nuclear power campaign in the country. Although I saw the DBA as NIMBY (Not in My Back Yard), Friends of the Earth's job was to convert NIMBYs into NAMBYS (Not in Anyone's Back Yard). Friends of the Earth at that time may have been a fringe organisation. But its backing of the DBA and hundreds of other similar groups has made the environment issue mainstream."

With his broad "green" outlook, Don also commented, "I remain pessimistic that members of organisations such as the DBA would be prepared to make the lifestyle changes needed to save the global environment. They want cars, electrical power and the material and financial success available to privileged groups in the late 20th century."

It seems to me that Don was justified in his views that NIMBYS can become NAMBYS. Despite protestations that they were not an anti-nuclear group, when I look back on old newspaper cuttings many of the statements made by spokespeople for the DBA in 1979 were highly critical of nuclear power. When I asked members about this, it was clear that as they found out the nature of what they were dealing with they found it impossible to sit on the fence and take no position for or against nuclear power in principle.

Pat Tudor explained how his views developed. "At first, we were local residents who needed, quickly, to begin to understand the arguments. If you live as we did in remote Northumberland, with no nuclear power stations around you, your stand on nuclear power is academic. At that time, people were not so conscious of environmental issues. I had never liked the idea of nuclear power, though I hadn't begun to understand all the arguments. But within the space of a year, I was standing on platforms arguing **against** nuclear power."

Alan Brown went through a similar process. He said, "My background was in engineering, with Reyrolle Parsons who made generators for power stations. But when I started looking into nuclear power, I became more and more against it. You find you have to go one way or the other."

It happened to Sue Jordan too. She said, "Finding out about nuclear power alarmed me. I didn't know a lot about the dangers before. There was never an answer to the problem of nuclear waste, and I don't suppose there ever will be."

• • •

The Druridge Bay Association began an intense programme of activities. One of its earliest aims was to ensure that Northumberland County Council opposed the nuclear proposal. DBA members were suspicious of the county council's fence-sitting position.

Pat Tudor said, "We were very concerned to find out what we could about the Druridge proposal, and to put pressure on our elected representatives to inform the local community about what we felt was a conspiracy to spring nuclear power on us. They may not have known more than we did, but we didn't know what they knew."

Another reason was outlined by Alan Brown. "We thought at the beginning that there was a limit to what we could realistically achieve. We set up the target of getting the council to oppose nuclear power at Druridge Bay. We knew that if we didn't get them on our side, we would get nowhere."

Colonel Jim Smail was chairman of Northumberland County Council's Planning Committee. He had come out with no clear statement of view on Druridge Bay. He had said in January 1979 that he was adamant that full public consultation would occur before any recommendation was made. By May, John Lodge,

County Planning Officer, reported that the CEGB were happy to support a programme of "public participation", which was therefore being arranged.

One may question the value of public participation. At least it brought CEGB people, the remote and elusive "them", out into the public eye. However it was clear that although allowed to "participate", no guarantee was given that the opinion of the public would be permitted to prevail.

In preparation for a series of public meetings, the CEGB had agreed to co-operate by supplying information to the people of the area. The county council therefore sent off a list of questions to the CEGB about nuclear power at Druridge Bay. Both the questions and the CEGB's replies were published by the County Council in July 1979 in a discussion document called *Nuclear Power Station Investigations Druridge Bay*. (From now on this document will be referred to as *Nuclear Power Investigations*.) Alan Brown was sceptical in principle of the value of information provided by the CEGB. He said, "It is rather like buying a second hand car and relying completely on the word of the salesman rather than getting an independent report."[5]

The Druridge Bay Association obtained a copy of *Nuclear Power Investigations* and put their heads together. None of them knew much about the subject, but they felt the document had not elicited the detailed answers they required from the CEGB. They took one month to prepare a 23 page document of their own, entitled *Nuclear Power at Druridge – Some Facts Against*. (From now on this document will be referred to as *Some Facts Against*.) In it they state, "Some questions have remained unanswered. Some answers are so general as to be unhelpful, and some are contradictory... There are many people who, although no experts in nuclear engineering nor electricity requirements, are not prepared to accept without criticism some of the misleading and superficial answers given by the CEGB to the County Council's questions."

Some Facts Against was impressive. From nowhere, the DBA had to come up with arguments, facts and figures. They were fortunate that one of their members was Peter Matthews, a scientist. He applied his skills to the CEGB's answers in *Nuclear Power Investigations* with help from other members of the DBA committee.

The CEGB's answers contained the first information that they gave out on the Druridge project. Their original press release which started all the furore in December 1978 had said simply that the Central Electricity Generating Board "wishes to investigate the Druridge Bay area in Northumberland to assess its potential as a nuclear power station site. Surveys of this area are expected to start early in the new year and would include exploratory drilling and hydrographic work. The Board has no firm plans for power station development in the north east."

In *Nuclear Power Investigations*, one can see the way the CEGB were thinking in

1979. It pointed out that a new, large nuclear power station would be needed for the north-east area because electricity demand would be likely to increase by 25% by 1990/91. To secure an adequate security of supply to north-east industries and homes, the CEGB considered that the best solution would be to build a new nuclear power station at Druridge Bay.[6]

In *Some Facts Against*, the DBA used information gleaned from the publication by Gerald Leach of the International Institute for Environment and Development, *A Low Energy Strategy for the United Kingdom*. They criticised the CEGB's forecasts as greatly exaggerated, and pointed out how electricity generating needs by 2000 AD could easily be met without a large expanding nuclear programme. Their criticisms have turned out to be broadly correct.

The DBA also calculated that Northumberland produces six times more electricity than it uses, and believed that any new power station should be sited in the Tyne/Wear/Tees area. This, if seen to be absolutely necessary, should be generated by combined heat and power, which uses fuels far more efficiently than conventional power stations.

Why was Druridge Bay chosen? The reasons given by the CEGB were that it has the right combination of technical and amenity requirements, which are:

- access to the sea for cooling water
- large areas of flat land not too far above sea level
- possible good subsoil conditions for support of heavy loads
- far away from built up areas
- outside the Area of Outstanding Natural Beauty which extends from Berwick to Amble.

The DBA pointed out that the CEGB were not investigating any other possible site between Lincolnshire and the Scottish border, and concluded that a decision to build at Druridge had already been made. They stated that 60 000 people live within 8 miles of the area, and questioned the stability of land riddled with mining and post-opencast land settlement.

The DBA also criticised the CEGB's safety analysis in some detail, and pointed out that the Three Mile Island reactor in the USA which nearly melted down in 1979 was similar to the one proposed for Druridge Bay. They poured scorn on the one kilometre evacuation zone around current existing nuclear power stations. The CEGB's position on evacuation was, "In all credible fault sequences, regardless of weather conditions, the release of radioactivity would not result in any need for action to be taken off the power station site."

Claims by the CEGB that 1500 – 2000 jobs would be available during construction were heavily criticised by the DBA. They pointed out that these jobs were largely unsuitable for local unemployed workers.

It is interesting to see that in 1979 the perennial arguments for and against nuclear power at Druridge Bay had appeared. They remain basically the same arguments to this day. They revolve round the local environment, whether or not the electricity is needed, safety issues and employment. Other perspectives have been added as further environmental issues emerged, but the early campaigners isolated the key issues.

The DBA booklet, *Some Facts Against*, was circulated widely. One of its primary uses was to give information to those councillors who were prepared to speak out against nuclear Druridge, and to present the opposition viewpoint to those who would otherwise have only seen the CEGB's information.

From the perspective of the early 90s, certain points made by the CEGB in *Nuclear Power Investigations* show their deviousness and wishful thinking.

One example is that they always planned to build **two** nuclear power stations at Druridge Bay, although they permitted the general idea to prevail that it would be one. They said, "The present studies are based on a possible initial power station development... and allow for a possible second phase of development."

In another example, they said, "It takes at least six years to build and put to work a power station which therefore requires construction to start in 1983/84 for 1990 commissioning." They tried to persuade us that the electricity was needed by 1990, giving us the feeling that construction was fearsomely immediate. Yet 1990 passed, and the need for this electricity has not materialised. We may be thankful that construction was not started when they thought it should.

Finally, the CEGB had an almost touching belief in the ability of the nuclear industry to dispose of old used-up nuclear power stations, in the light of what we now know about the difficulties and costs of decommissioning. They said, "The reactor structure itself, which is radioactive, **cannot be dismantled for some time.** The nature of the radioactivity associated with a reactor structure would make it advantageous to **delay its dismantling for some 50 or more years after shutdown. Studies are being carried out** to determine at what stage final dismantling might take place." (author's emphasis) Clearly, local people were expected to trust the CEGB that the "studies" would come up with an answer, and to accept blithely a delay of at least 50 years before decommissioning of the radioactive hulk.

• • •

The documents embodying the key arguments from both the CEGB and the Druridge Bay Association were ready by September 1979. The consultation exercise, which was to lead to a decision by Northumberland County Council on its position on Druridge Bay, was set in motion.

The first public meeting was the Ashington seminar, on 3 September. Invited to be on the platform at this meeting at Ashington Technical College (now

Northumberland College of Arts and Technology) were two representatives from the CEGB, one from the Nuclear Installations Inspectorate, and one from Northern Engineering Industries Ltd. The opposition was represented by Mike Flood from Friends of the Earth and independent statistician John Urquhart. Mr J Parker came from the National Coal Board, whose position was ambiguous. The forum was chaired by Professor G R Bainbridge, Director of the Energy Centre, University of Newcastle.

Five of the eight items/speeches on the agenda were in favour of nuclear power. Two were against, one was unclear.

The seminar went on from 10.00 am till 4.30 pm. The speakers put forward their views for or against nuclear power in general and at Druridge Bay in particular. Local reporters attended and gave detailed coverage. Despite the heavy balance of speakers in favour of nuclear power, there was no trouble at the meeting from opponents until the summary at 3 pm.

Professor Bainbridge chaired the Open Forum, at the end of which he was to comment on and summarise the issues raised at the seminar. However Druridge Bay Association members were outraged by what they felt was his biased summary, and this led to a dramatic end to the meeting.

Pat Tudor well remembers that day. He said, "I stood up and cricitised the chairman's summary as a total distortion and misrepresentation of the events of the day. There was discussion and some recognition of this, and almost an apology from the chair. Then he went on to do it again. I said something like this. 'I've already interrupted once Mr Chairman, but I'm aggrieved that no notice has been taken of my representation. I am so frustrated that I want to invite all those people who feel this is an utter distortion to join me by walking out.' And a substantial number of the gathering walked out."

Wansbeck District Councillor Leo Kemmit was one of the people who walked out. As he left he said, "Everyone is entitled to their view, but I think it is the job of an independent chairman to remain neutral. I don't think his brief was to come down strongly the way he did. If he had wanted to do that he should have been in the body of the hall with the rest of us, or had himself appointed as a speaker."[7]

Even Colonel Smail felt he had gone too far. He said, "The chairman is entitled to his views although we were not aware they were to be given. But it must not be said that they were the views of the county council; we came to this teach-in to learn and not to come to any conclusions."[8]

Pat Tudor said, "I knew when I walked out I would carry a dozen or so people with me, people whom I knew. But I didn't realise how strongly others would feel. Later, I was worried that I would get into trouble at work. My name was mentioned in at least two local newspapers as leading the walkout."

Alan Brown had even stronger feelings about the seminar. "It became clear to us that there could be collusion. We felt that certain attenders were very positive towards the CEGB proposals, almost to the point of silliness."

However, the participation process was in motion, and the Druridge Bay Association did its best to encourage public opposition to make itself heard. Following the Ashington seminar were the three public meetings in Broomhill, Widdrington Station and Ashington.

I asked Alan Brown if he felt the meetings were organised fairly. He said, "We felt the county council had to be seen to be doing it. It was a face-saving exercise, so they could say they'd had public consultation.

"We were annoyed that they had their public meetings with a member of the CEGB on the platform, while we could only question from the floor. However we were a real thorn at those meetings. We wanted to make sure all our arguments were presented. We went in force as a committee. We put the same embarrassing questions each time."

Obviously DBA members felt convinced that the CEGB had power and influence far greater than their own. All they could do as individual members of the public was to question the powers-that-be, provoke them, and encourage the other people to challenge their seemingly immoveable authority. People needed to have faith in their own convictions that nuclear power at Druridge Bay was totally unacceptable, and have the information to back their position.

The Druridge Bay Association was relieved by the reaction of people who packed the public meetings. Following them, Alan Brown stated his view that plans for a nuclear power station at Druridge were roundly condemned, and that what had emerged was unanimous opposition.[9]

Having heard the opinions of local people, the county council retired to consider its position, which was expected to be announced at the end of the year. Then in December 1979, once again just before Christmas, the government made an announcement which increased local anxiety about Druridge Bay. Energy Minister David Howell announced that there would be a 2 billion programme over the next ten years, building one new nuclear power station a year. The power stations were likely to be the pressurised water reactor (PWR) of the type which nearly caused a disaster at Harrisburg in the United States earlier in the year. This was a new technology for Britain.

The ten new sites were not named, except for the first which would be at Sizewell, in Suffolk. However local people realised that Druridge Bay was certain to be one of them. The Druridge Bay Association knew that the battle over the pressurised water reactor would probably be fought at Sizewell, at a public inquiry. This would give people fighting for Druridge Bay a breathing space.

The announcement of the new nuclear sites certainly brought the nuclear power issue to the forefront of people's minds at the very time when the county council was considering its position. Despite an instinctive preference for nuclear power among some councillors in this Conservative/Independent council, there was still a need to take into account the electors' views.

The outcome of these meetings and internal deliberations at the county council did not materialise until February 1980. After some re-drafting of its position, Northumberland County Council took a stand opposing nuclear power at Druridge Bay. The council agreed that a power station at Druridge Bay would be "contrary to local planning policy concerning the environment, agriculture and recreational activity," and "On the information now available, the Planning Committee are not convinced that a case has been established by the CEGB to cause them to depart from their planning policy."

The Druridge Bay Association was delighted. Alan Brown, as chairman of the DBA wrote to members in the March 1980 newsletter, "Round 1 to the Druridge Bay Association and the people of Northumberland. I believe the the DBA and the work which it has carried out, television interviews and radio broadcasts, local and national, public meetings, and hundreds of newspaper reports, has been primarily responsible for alerting and eventually convincing our County Councillors. It should be remembered that it was not until well after we started to pressurise the County Council that their public consultation exercise was initiated."

Councillor Jim Smail, chairman of the planning committee, said that Alan Brown's claims were an insult. "Mr Brown is being conceited to suggest this, and it is just not true to say he pressurised any members of my committee."

As far as the Druridge Bay Association was concerned, one of its principle aims was achieved by the County Council's statement of opposition to nuclear power at Druridge Bay. It had taken from December 1978 to February 1980 to reach this point.

The pressure group certainly felt it had contributed to the decision, though such a matter is hard to prove. While monitoring the County Council's decision-making process during 1979, the Druridge Bay Association had been involved in a hectic year of activities, acting in all the ways typical of a pressure group which intends to cultivate public awareness. These included letter writing to councillors, speaking at public meetings, setting up and maintaining membership schemes, linking up with groups opposed to nuclear power nationally and locally, organising events, taking part in demonstrations, and taking an information caravan around local exhibitions and shows.

This work had been shared among the committee, with support from members. There were never enough people to do all the work, as always seems to be the case with voluntary organisations, yet it still went on.

On Saturday 7 July, principally due to the initiative of Harry Allport, there was a large rally and musical event at Druridge Bay. Over 3000 people flocked to the sand dunes, on land owned by farmer Derek Storey.

Dennis Murphy, president of the Northumberland miners, spoke at the rally. He said, "If this power station is as safe as they say, let them build it in Battersea to prove it." He claimed there was enough coal under the site to provide electricity for fifty years, and added, "There are seams under the sea which could be mined, leaving no disfigurement to the land."[10]

Other speakers at the rally included Dr Gordon Adam MEP and Councillor George Loggie, Chairman of Wansbeck District Council.

Shortly after the rally, the reality of the nuclear proposal was brought home as drilling rigs were moved by the CEGB on to two sites at Druridge Bay during August. The sites were at Hemscott Hill near Druridge hamlet and Whitefield Bank Farm near Red Row. This kept emotions at fever pitch.

The Druridge Bay Association objected furiously to the chosen date of moving in drilling equipment. Sue Jordan said, "We were told the drilling would not begin until after harvest time, but they have sneaked in without any consultation with residents." She also said that it was suspicious that the electricity board had

Rally at Druridge Bay organised by the Druridge Bay Association, 7 7 79. *Photo Newcastle Evening Chronicle and Journal Picture Library.*

chosen a miners' holiday to start their test drilling, as the miners had threatened to picket the test site. The CEGB however denied there was any plot in the timing, claiming they had waited until the county council's Discussion Document was published before tests were carried out.[11]

The drilling rigs got on with their jobs during August and September, while the County Council's consultation exercise was going on and the Druridge Bay Association kept up its activities. However, the target provided by the rigs proved too tempting for some unknown opponents, who undertook a little direct action.

Alf Townley reported how he first found out that there had been sabotage. Alf lived at Blakemoor farm, next door to Gordon Bell, who farmed Hemscott Hill where the drilling was being done. "Gordon Bell told me that someone had cut the pipes on the drilling rigs. The press contacted me, and I told them to ask Gordon Bell, but he would not comment. Whoever did it knew what they were doing. It was discovered when they tried to start drilling. The newspapers only gave it a very brief mention. I think the story was stopped."

According to the *Journal* and *Evening Chronicle*, who both ran a brief article on the sabotage, there were two incidents. The first involved sand and soil added to the fuel and hydraulic tanks, and lights on the rig were smashed. The second event involved sugar being added to the fuel tank. Norwest Holst Ltd who were carrying out the drilling said they had increased their two man security team, and walkie-talkies and dogs were being used to maintain a 24 hour watch.

The DBA disassociated themselves from the sabotage. At the time, Sue Jordan told the *Journal*, "This has not got the blessing of our association. We want peaceful action." She said that the DBA had no idea who had done it, but added, "It could have been anyone in the area, the feeling is so strong."

The drilling went on until the end of September, and this was followed by the miners' rally in Ashington. The event was supported by the Northumberland and Durham miners, the DBA, Friends of the Earth and Wansbeck District Council. Sam Scott, secretary of the Northumberland NUM declared, "Nuclear power will mean loss of jobs in our northern coalfields and it is a potential danger to the health of the people." These speakers, and Gerard Grenville of the Society for Environmental Improvement, all agreed that nuclear power was the stick with which the Conservative government meant to beat the miners into submission.

Many shades of opinion were represented at this rally. Alan Brown, as a leading member of the DBA which had set itself up as a residents' association, commented, "I never in my life imagined I'd lead a march with Arthur Scargill. But there I was, walking with him, speaking to him. That's how your life changes."

From its earliest days, diverse groups were opposed to nuclear power at

Druridge Bay Association Demonstration in Ashington, 1979.

Druridge Bay. There are many reasons for opposing nuclear power on a coastal beauty site, on agricultural land. In October, the National Trust declared, "A large new industrial complex so very close to this beautiful coast, with its great amenity value, must be opposed in the absence of much more information about the shortcomings of alternative and less sensitive sites that have presumably been investigated."[12]

There were in 1979 some efforts to initiate collective action, among groups as widely different from each other as the Druridge Bay Association and the National Union of Mineworkers. In August, a three hour meeting in Ashington had been attended by the NUM, the DBA, the National Farmers Union, Anti-Nuclear League, Tyneside Environmental Concern, Friends of the Earth and local trades councils. Alan Brown commented at the time that cash to mount a major campaign had been a headache, but following this meeting, pooled resources were now available.

The umbrella group had another meeting in September, when the name NADBANG was coined. This stood for Northumberland and Druridge Bay Anti-Nuclear Group. However this federation was never properly consolidated. It can take the credit for being the first attempt to unite all shades of opposition in a sturdy federation. The idea was sown. It was not however until the formation of the new Druridge Bay Campaign in 1984 that it blossomed into reality.

A major publicity event organised by the DBA in October was the showing of the film *The China Syndrome* in Morpeth. An anonymous supporter of the DBA donated the cost of hiring the Coliseum cinema in Morpeth for a special private showing. In the audience were county and district councillors, representatives of Northumberland Area Health Authority, the NUM, and Wansbeck Trades Council. Among those present was Jack Thompson, at that time leader of the Labour opposition group on Northumberland County Council. He later became leader of the council, and then MP for Wansbeck. Commenting on the film, he said, "It has substantiated my views. I started off against the Druridge Bay proposals, but I was prepared to be convinced otherwise through the whole exercise of seminars etc we are going through." Jack Thompson's continuing opposition to nuclear power at Druridge Bay has been a decade long source of strength to Druridge campaigners.

Other councillors made similar comments, including Jimmy Nicholson and Alan Thompson. Councillor Jim Smail, who was unable to attend, said, "I am quite sure it is a thrilling film, and I am hoping to see it in London before the county planning committee makes a decision. When you are making planning decisions you cannot judge on a fictitious film. The planning committee will make a common sense approach."[13]

Shortly after the showing of the *China Syndrome*, Morpeth Civic Society heard a DBA member speak on the Druridge issue. It was only one of numerous speaking engagements the association undertook, but important to me because that was when my husband and I first became involved. We were members of the Civic Society, and had asked for a speaker on Druridge Bay to be invited. It was our first encounter with the DBA, and we were very favourably impressed. Alan Tipping, DBA treasurer, went over the arguments in a calm, reasonable manner. Afterwards, my husband and I paid our fifty pences to become members. Besides signing the petition outside Laws Stores in Morpeth, it was the first positive action we had taken for Druridge Bay.

The continuous series of actions organised like this by the Druridge Bay Association during 1979 must certainly have prompted the public's response, and kept the county council on its toes. By articulating the opposition viewpoint, in particular by the production of the booklet, *Some Facts Against*, the DBA challenged the information provided by the CEGB.

The association worked on beyond 1979, but this was its most hectic year.

Looking back on that year a decade hence, Alan Brown said, "It was a good time but it was amazing how much it took out of us. I was working for a small electronics company as general manager, with a view to becoming a third director. It was a good job. But the Druridge proposal came up almost immediately after I took the job, and was the reason why I left in the end. My employers were very supportive, but it was getting crazy. There were constant phone calls from newspapers, TV and radio, wanting comments. At times the owner of the company used to get extremely irritated. I left there after one year and ten months."

After the Secretary of State for Energy's statement that ten new PWRs were to be built in the next decade and the declaration of opposition by Northumberland County Council in February 1980, there was a gap in activities. Nothing very much happened about Druridge Bay during the rest of 1980. There were items reported in the newspaper about the increasing awareness of Northumberland County Council that coal was the way to generate electricity for the North East, and that a site available at Blyth next to the two existing power stations should be developed. There was intense activity slightly further north in Northumberland about drilling for nuclear waste in the Cheviots. That is a story in itself. But there was little that seemed to be happening about Druridge Bay.

The Druridge Bay Association kept up as best it could its numerous activities, but the stimulus was declining once Northumberland County Council had taken its stand and no developments were forthcoming from the CEGB.

As with Torness Inquiry, there can be seductive lulls in activity with the nuclear industry. Campaigns find it hard to sustain themselves when there is no visible enemy. People drift off into other activities. It is dangerous however to be seduced like this. The bombshell was delayed, but it came at last.

It was another Christmas shock, in a pattern that was becoming familiar. Christmas 1978 had brought with it the original announcement of the CEGB's interest in Druridge Bay. Christmas 1979 was accompanied by the government announcement that there would be ten new nuclear power stations for a decade. Now Christmas 1980 brought the worst possible news. The drilling had proved that the site at Druridge was suitable for 2500 Megawatts of plant which could be installed in stages. Once again, this was CEGB-speak for two large nuclear power stations.

Two years after horrifying the local population, the CEGB confirmed their interest. The land was suitable. There was no more room for hope that this would not be the case. The Hemscott Hill site had been chosen. The CEGB declared that there were no immediate plans to put in a planning application, or decision on which reactor type would be used.

There would clearly be a long struggle ahead. In the 1979 *Nuclear Power Investigations* document, the CEGB had stated that construction would need to start in 1983/84 for 1990 commissioning. In December 1980, Mr Len Nash of the CEGB accepted that demand for electricity had reduced, but this only meant putting back the date when a new power station would be needed.[14]

At the elections in May 1981, the Labour Party took control of Northumberland County Council. Jack Thompson became leader. In February he had stated, "Our Labour group is definitely against a nuclear power station at Druridge Bay. We would only accept as a possible future development a coal-fired power station at

Blyth." At the helm of the county council was now a group unequivocally opposed to nuclear power at Druridge Bay. It consolidated the growing awareness of the previous administration that the answer for the region lay in coal, for reasons of environment, safety, and local employment. A coal fired power station at Blyth would be an acceptable alternative to nuclear power at Druridge Bay for Northumberland County Council.

Still the months passed by with no definite moves. By 1982, the Sizewell Inquiry was looming. It was clear that a planning application for a new nuclear power station in Suffolk was going to be the focus for the whole nuclear power debate in England and Wales. Once a decision had been taken in principle on the introduction of the PWR technology to Britain, further public inquiries would be limited to local concerns. Because of this, north-east councils had to decide whether or not to make representations at Sizewell on wider principles. Northumberland County and Wansbeck agreed that they would do so. Castle Morpeth decided against it. (The Sizewell Inquiry in fact did not open until February 1983.)

Another bombshell was dropped on the area when, in August 1982, Druridge Bay was confirmed as being on the shortlist for new nuclear power stations. The next site proposed after Sizewell was at Hinkley Point in Somerset. After that, a short list of four sites for new nuclear power stations included Dungeness in Kent, Sizewell again, Winfrith in Dorset and Druridge Bay. Six of the ten nuclear power stations mentioned by David Howells in late 1979 had been named.

Despite the forthcoming Sizewell Inquiry, an immediate, visual target for the Druridge Bay Association was lacking. The strain on the personal lives of the main activists was hard for them to cope with. By August 1982, after the announcement that Druridge was on the short list, Sue Jordan responded to a *Journal* reporter that new blood was desperately needed by the Druridge Bay Association. Now described as ex-secretary of the DBA, she appealed for interested people to make their views felt by joining the group. "People have been ringing me up since the latest news was announced and asking what they can do," she said.[15]

Looking back on that time, she said recently, "It was difficult after the first couple of years, after the initial surge of energy which went into the campaign. Most of us were working full-time, and travelling some distance to work. I had started a family. It was hard to keep up the momentum."

Other members of the DBA committee found that the lack of action by the CEGB, combined with the demands of the their own personal lives, added to the tapering away of their organisation.

Pat Tudor said, "Where does the momentum come from for a campaign to save Druridge Bay? In part it came from the CEGB. They acted. We responded. How could we retain the thread of continuity in the affairs of the organisation in a place where perhaps twenty per cent of the population move house in a ten year span?"

Alan Brown said, "We knew it would be a long haul. The issue may not even be resolved in our lifetimes. But it is a problem once you let your guard down. The CEGB had the resources to keep going. We knew by reacting early, our strength would be sapped, whereas their people are paid to do the job. We all got tired. We moved. We had a life to lead. We were aware of that, but it was hard to keep going."

In the years 1979, 1980 and 1981, the Druridge Bay Association had been the key pressure group. Although councils, trades unions and other organisations developed their opposition within their respective frameworks, the DBA existed for no other reason than to oppose the nuclear proposal despite its much wider constitution.

The opposition from the DBA and other groups, had remained unco-ordinated. Vince Gledhill, the *Evening Chronicle* reporter based at Ashington, followed the Druridge story for many years. Looking back to the early days, he described the fragmented nature of the opposition.

"Opposition to the proposal was strong, but unco-ordinated, and did little to shake the confidence of the bureaucrats. Some organisations such as local councils and unions opposed the plan, but it was difficult to divorce how far they represented wider public interest from their own secular interests or from entrenched political positions. Another polarised position was that of people objecting to the plan on the grounds that they did not want a nuclear power station on their doorstep. Inevitably, the more fragmented were these strands of opposition, the easier it became for supporters of the power plant to argue that only minority interest groups were against it."

The DBA deserves much credit for setting the pressure group process in motion. The miners, the councils and other groups were making crucially important contributions. An attempt to form a federation had been made, but had fallen through. It was to be a long struggle, and there would be time in the future for the campaign to develop.

Ian Breach was producer of the TV programme Grapevine, which ran an item on the Druridge Bay Association. Sue Jordan, Alan Tipping and Alan Brown were filmed in their homes, and canvassing local residents. At the end of the programme, Ian Breach summarised the role of the DBA.

He said, "A small group, initially possessing very little knowledge of a technical and complex subject, can quickly develop its expertise, can muster the skills to lobby effectively, and can soon acquire the confidence necessary to challenge even major institutions like the generating boards and the Atomic Energy Authority. The message is a clear one. However modest your resources, however large the organisation against which you are pitting those resources, you can have an effect. You can join the debate."

NOTES

1. *Morpeth Herald*, 19 1 79
2. *Morpeth Herald*, 5 1 79
3. *Morpeth Herald*, 16 2 79
4. *Morpeth Herald*, 2 3 79
5. *Evening Chronicle*, 29 8 79
6. In *Nuclear Power Investigations*, the CEGB gave information for two regions; the North Eastern Region which stretches from Humberside and South Yorkshire to the Scottish Border; the North East Coast Area which includes Teeside, Wearside, Tyneside and Northumberland north to the Scottish Border.

The forecast increase in demand for the North Eastern Region was from 7330 MW in 1979/80 to 9300 MW by 1990/91, an increase of 25%.

The CEGB stated a large new nuclear power station would be needed for the North East Coast Area because electricity demand would be likely to increase from 2380 MW to 3200MW by 1990/91. In 1979 local coal fired power stations were providing about 2400 MW. However, the expected increase had to be provided for. At times of peak power, extra electricity was already being transmitted from Yorkshire and Humberside power stations, and the CEGB claimed that these forecast **additional** requirements could rise to 900 MW by 1990/91. To ensure adequate security of supplies to North East industries and homes, either a new power station had to be built in the area, or a major new transmission interconnector to the south installed and the new power station built elsewhere.

We now know that the CEGB vastly overestimated electricity needs. Their forecast for the whole North East Region by 1991, which includes the North East Coast Region described above, was as follows:

High Forecast, 10 700 MW;
Medium Forecast, 9 300 MW;
Low Forecast, 7 800 MW.

Actual electricity consumption for 1989/90, the latest year for which figures are available at time of print, was **7300 MW**. This is the same as consumption in 1979. In other words, there has been no increase in demand.

7. *Northumberland Gazette*, 7 9 79
8. *Northumberland Gazette*, 7 9 79
9. *Evening Chronicle*, 14 9 79
10. *Sunday Sun*, 8 7 79
11. *Journal*, 6/7 8 79
12. *Evening Chronicle*, 11 10 79
13. *Journal*, 29 10 79
14. *Northumberland Gazette*, 5 12 80
15. *Journal*, 31 8 82

2. FRIENDS OF THE EARTH MID-NORTHUMBERLAND
1983
Reviving the opposition

By early 1983, Druridge Bay was no longer in the news. The Druridge Bay Association was quiet, and like many other local people, I was beginning to wonder what was happening about the nuclear power proposals.

My own and my family's involvement in the issue began more or less by chance in this era of no activity. One day, in February 1983, a tiny notice appeared in the *Morpeth Herald*, announcing that Durham Friends of the Earth were holding the results of a draw at Druridge Bay. The draw was one of their activities to raise money for Friends of the Earth to fight at the Sizewell Inquiry. The result of this Inquiry was most important to Druridge Bay, as it was to debate whether a new generation of American nuclear power stations, the pressurised water reactor, (PWR) should be introduced into Britain. One of these would be at Druridge Bay.

My husband and I decided to go and see what was going on. We drove out to the *Widdrington Inn* with our children, Patrick aged 12, Daniel aged 9, Jeannie aged 5 and Laura aged 3. We were vaguely expecting crowds of protestors and banners. A few cars were parked outside the inn. One or two people who might have been Friends of the Earth were walking around, but we were not at all sure what such people looked like.

Then we saw some friends of ours from Morpeth pulling into the car park, with their children. Soon there was another car. The Stacys and the Conways had seen the same newspaper article, and had been drawn by interest like ourselves.

We identified the Friends of the Earth members, who explained that there were only a few of them. They had come from Durham to Druridge Bay to make a link between the North East site and Sizewell in Suffolk. We met Don Kent, a big jovial man with a white plaster on his broken nose, who encouraged us to buy tickets for the draw.

Our son Daniel, who is always lucky in draws, won a Save-the-Whale T-shirt, and £5.00. We chatted with the Durham people, and it was only a small step to realise that with their encouragement we should start a Friends of the Earth Group locally to raise the Druridge issue again.

The *Courier* editor, Jim Mathieson, arranged us all outside outside the pub with such posters as the Friends of the Earth had brought, took our photo, and a new phase of opposition to nuclear power at Druridge Bay had begun.

Lindy Conway, Rosie Stacy and I soon met at one of our houses, and made plans. We telephoned Sue Jordan, who explained how the Druridge Bay Association had become totally dormant. We decided to make posters for a meeting to be held on the Druridge issue, under a Friends of the Earth banner, in Morpeth. We put the posters round the town, and hoped people would come.

Don Kent came to the inaugural meeting. He spoke about Friends of the Earth's general activities, encouraging the group not to concentrate solely on Druridge Bay. The fifteen people who came were however much more motivated for that cause than any other. *The Evening Chronicle* announced our group with big headlines, DRURIDGE BAY BATTLE LINES. If it was as easy as that to get in the papers, we had made a start.

Immediately, we decided on a programme of activities. We set up a petition, and sold balloons for a balloon race from Sellafield in support of a fund-raising exercise organised by Don Kent's Tyneside FoE. Mainly, however, we wanted to focus on Druridge Bay, and with something more visual than merely a petition. We came up with the idea of building a cairn at the Druridge nuclear site.

One of our members Ed Metcalfe comments, "The idea of building the Cairn came out of a brainstorming at a Friends of the Earth meeting. It grew on us during the session. It was simple, and all good ideas have to be simple. It was universal, not political in any way. It was on the site, local. It had a 'green' image, an image of being associated with the land. It was organic; people could participate in its growing. If we had just stuck a big sign up, people would just look at it and wander off. But people made a point of bringing a stone to put on the Cairn."

The Cairn symbolised also the simplicity and powerlessness of ordinary people. When we announced our plans to build a Cairn, the Newcastle *Evening Chronicle's* headlines were A STONE PILE TO FIGHT NUCLEAR ONE.

It was my job to invite the local VIPs to the founding of the Cairn. When later we were accused of being "professional protestors" by Fred Fever of the CEGB in the *Northumberland Gazette*, I think back to those days. We knew very few VIPs between us and I was so nervous at having to ring up such important people as headmasters, or Jack Thompson who was then leader of Northumberland County Council, that I had to force myself to cross the room to make the telephone calls. I well remember Jack Thompson's kind response, although he explained he would be on holiday on our chosen date. We were thrilled when so important a person as Alan Beith MP was free and willing to come.

Mr Storey, the farmer who owned the land immediately adjoining the site in which the CEGB were interested, kindly agreed to let us borrow the corner of his field near the National Trust entrance. It was an ideal site, and when the Cairn grew bigger, could easily be seen by passers-by over the wall.

It may appear odd, but one thing sandy Druridge Bay does not have is stones. The problem of finding a load of stones and taking them to Druridge Bay had to

The first stones of the Cairn at Druridge Bay, 30 3 83. *Photo Newcastle Evening Chronicle and Journal Picture Library*

be solved. Fortunately, Councillor Gordon Knox of Blyth Valley Borough Council came to our aid. He arranged for a load of demolition stone from a site in the Blyth area to be made available. Wansbeck District Council agreed to collect the stones and deliver them to Druridge Bay.

On a bleak March morning in 1983, about 40 people gathered at the Cairn site. We had arranged the stones in a large open circle, and people assembled behind them. Alan Beith stood in the middle of the circle, and spoke about the threat to Druridge Bay. Then he picked up the first stone, and placed it in the middle of the circle. Other participants followed. A little cairn about a metre high was built. There it was, small, but ready to grow; a visible symbol of people's feelings.

Those who attended the ceremony symbolised the variety of views that would continue to oppose nuclear power at Druridge Bay. As well as Alan Beith MP and Friends of the Earth members, the following people came:

from Castle Morpeth Borough Council
Cllr Matthew George Green
Cllr Mrs Wendy Sayer
Cllr John Lough
Cllr Mrs M Smith
Cllr Humphrey Devereux (also Northumberland County Council) and family

from Northumberland County Council
Cllr Mike Stott and family
Cllr Gordon Knox (also Blyth Valley Borough Council)

from Wansbeck District Council
Cllr J Devon (Chairman)
Cllr D Scott (Leader)

from Blyth Valley Borough Council
Cllr Ian Gordon (Leader) and family

from Ellington Parish Council
Cllr R Scott

from Alnwick Town Council
Cllr F Williams
Cllr T J Howells
Cllr A J Gordon

Mrs Pat Partington, Mayor of Amble
Mr Partington
Professor Mike Clark, Newcastle University
Dr Marjorie Mowlam, Newcastle University, now MP Redcar
Roy Beasley, Principal Ashington Technical College
Mrs Anne Swailes, teacher
Mr Alec Swailes, teacher
Mr Roland Bibby, historian
Mr Lonsdale, National Union of Mineworkers
Mr Michael Duffy, Headmaster King Edward V1 high school, Morpeth
Mrs Pauline Duffy

David Rhodes donated the explanatory sign on the wall.

Some of the people who attended the founding of the Cairn ceremony told why they came and how they felt about it.

Roy Beasley, Principal of Ashington Technical College, said, "I went with my family because my wife and I felt strongly about the potential health hazards associated with the proposed nuclear power station, because of its effect on the precious local environment, and because of available alternative power sources. The recent findings regarding transmitted cancer in children of Sellafield workers and the circumstantial evidence of leukaemia clusters there and near Dounreay and the other nuclear sites seem more than adequate justification for our concern as parents of children living in the vicinity of Druridge Bay. The unsolved problems of waste treatment and disposal, as demonstrated by the proven pollution in the Irish Sea, provided further justification, if any is needed, on environmental grounds. The Cairn ceremony was a focus for an optimistic and determined group of people from all walks of life who would provide formidable opposition to Government agencies and vested interests that might try to steamroller the nuclear proposals through the system."

Alec Swailes, local headmaster, singer and performer, was born in Newbiggin in 1932. As a child, he was frequently taken to Druridge Bay on church or family outings.

He said, "I have watched the dreadful intrusions up the coast for a lifetime. There was no power station at Blyth in my youth. I had to go to the Cairn founding ceremony to stop this creeping monster of industrialisation and danger. I have been a park warden and frequent walker in the hills for eighteen years. A cairn marks the pinnacle of achievement, and a guide in stormy weather. That is its symbolism for me."

Michael Duffy, headmaster of King Edward V1 High School in Morpeth, said, "My wife and I went to the Cairn ceremony because of our concern for Druridge Bay itself. Unspoilt coastline is a rapidly diminishing asset in the country. We would oppose any industrial development on it. The Cairn was a small symbol of a big threat, What interested us – and encouraged us – about the event itself, was the degree of cross-party support it represented."

The Cairn was attended that summer, and has been every summer to date, by campaigners with leaflets and the petition. Friends of the Earth member Hilda Helling described it. "It has become a landmark. Not only is it a physical reminder to all who see it of the strength of opposition, but a spiritual signpost of hope and encouragement."

Blyth Valley Borough Council ran out of demolition stone after supplying us with the first couple of loads. They offered us granite paving stones. I was walking towards the Cairn when the delivery truck passed me. It arrived at the Cairn, and I saw to my dismay the truck tip up and unload enormous paving stones which must have weighed a couple of hundredweight each. It was too late to do anything, and the truck drove off.

The stones were so massive an adult man could not lift them. Ed Metcalfe remembers, "They were very heavy. We tried to break them into manageable pieces with a sledgehammer. It was jolly hard work. Lots of little sharp bits flew around. We ended up being quite happy if we could get them just small enough to carry."

They were hauled over the wall, and gradually used. Some of them were left intact and later used to make a series of steps up to the top of the Cairn.

Now, in 1991, the Cairn stands four metres high, and is a sturdy pyramid. The only reason it is not higher is the problem of getting stones to Druridge Bay.

• • •

An early decision to be made by Mid Northumberland Friends of the Earth was what style of operations we should adopt. We also had to learn about nuclear power. It was like re-inventing the wheel. Very few of us knew in detail of the information and activities of the Druridge Bay Association. We were typical of the general population in our ignorance of nuclear power. We knew that there had nearly been an accident at Three Mile Island, and that nuclear power was seen as an alternative to coal for producing electricity. However the Chernobyl accident had not happened, Sellafield's famous leaks were yet to come, and the link between radiation and leukaemia scarcely heard of.

We had to respond to the energy debate as it existed then. The pending Sizewell Inquiry was to recommend on whether or not 1200 MW capacity PWRs should be introduced to Britain. These large stations are equivalent in size to Blyth B power station in Northumberland. The government was hoping to build up to twelve.

Also looming was the miners' strike. The miners feared that nuclear power was a threat to coal, and their livelihoods. The big expansion planned by the government through the PWR programme was an important cause of the strike. The miners' strike split the general population down political lines. To support the miners was interpreted as being left wing, as was being "anti-nuclear". To support nuclear power was to support the Conservative Party. Feelings ran very strongly locally as well as nationally.

As with the earlier DBA, the issue of whether or not to be anti-nuclear was to be settled. After discussion, the group agreed not to take a party political position, nor to be outright anti-nuclear.

Lindy and Paddy Conway, co-ordinator and treasurer of the group, put it like this. "It was always a political issue which would have a political answer. We used political means, too, right from the beginning. But we certainly decided that we would not ally ourselves to any political party, nor would we campaign against nuclear power as such. We wanted to demonstrate clearly that so many different sections and groups of people in the area were anti-nuclear-power at Druridge, and it was not a one argument campaign."

Ed Metcalfe commented, "To have operated within a political framework would have been counterproductive. It would have given the newspapers and the government something to put us down with. Most people were against the power station for amenity reasons. We had to make sure we didn't miss any of the support by working on as broad a front as possible. To be party political would have been failing to give a voice to the majority of people who were against the power station."

We felt it important to draw in people from all points of view across the political spectrum, to break the divide between left and right with respect to Druridge Bay.

A conversation in 1984 with the Mayor of Morpeth, Cllr George Green, epitomised the problem. When I asked him to attend an event Friends of the Earth were organising, he said he didn't want to be there with "all those lefty people waving banners around". I tried to make the point that that was exactly why he was needed. It was vital that opposition to nuclear power at Druridge was representative of all of the people of the area, regardless of their political beliefs and personal style.

Ed Metcalfe elaborated on this point. "We went to great lengths to get Conservative politicans involved. Berwick Conservative Association was supportive. Every time we contacted the MPs, we made sure we wrote to everybody, including the region's three Conservative MPs, Neville Trotter, Geoffrey Rippon and Piers Merchant.

"It was fortunate that the two local MPs who were keen to support the campaign, Alan Beith and Jack Thompson, were Liberal and Labour. With them, and tacit support from some Conservatives, we covered the main parties."

Also, the Greenham Common debate, and the forthcoming arrival of the American Cruise Missiles, was prominent. To be female, and "anti-nuclear", was tantamount to being a "Greenham Common woman". In fact, that was what my sons' friends used to call me at their school, even though I had never been near Greenham.

Rosie Stacy, who had been to Greenham Common, and supported the women there, explained why the Friends of the Earth's efforts had to be kept separate from the nuclear weapons debate.

She said, "If Mid Northumberland Friends of the Earth had been closely associated with CND, there would have been a great many who would not have joined it, people who would have supported saving Druridge Bay on environmental grounds. It was a pragmatic decision, and the right one. I think people become involved in protest movements gradually. They come in over a particular thing if it is in their back yard, like the threat at Druridge Bay."

Public opinion has changed a great deal over the last ten years about nuclear power. We did not anticipate this at that time. Although it was apparent to us that nuclear power, especially at such a place as Druridge Bay, was a monstrosity, the majority of the British public did not yet share that view. At first, we were not sure what the feelings of North East people would be. We soon found out.

Lindy Conway explained what happened when we were out campaigning.

"My impression was of overwhelming support. We got very few people coming up and arguing for the power station, or making rude remarks. An attitude that came to the fore was the feeling of hopelessness and powerlessness, the feeling that 'its all sewn up in London, and nobody will listen to us.' This was epitomised by one of the police at an event we held who said he totally agreed with us but would not sign the petition because it was hopeless."

Ed Metcalfe remembered, "We did get negative reactions from some people. A few refused to speak to us. But it was a very small percentage, and they often had a declared interest; for example, a man who said it would be good for business, or a woman whose husband worked at Parsons.

"I hadn't expected it to be so strongly favourable. People weren't just saying 'yes' to get rid of us. They were actually letting off steam, really getting quite upset about the whole thing."

Already it was becoming obvious that in order to succeed, the campaign to save Druridge Bay must cross political boundaries. It must be a campaign for local people, with a variety of interests. People could oppose nuclear power at Druridge Bay for their love of the bay, or fears for safety, or fears for their jobs and communities.

Within this social and political background, the newly fledged Friends of the Earth group started to operate. We had no power or influence, and only £5.00 in funds from the winnings of Daniel's draw. We had very little experience in public matters, and Rosie Stacy, Lindy Conway and I had been full-time mothers for some years.

• • •

As the founders of the Mid-Northumberland Friends of the Earth lived in Morpeth, the group tended to be Morpeth based. Jane Gifford, of Tyneside Anti-Nuclear Campaign, who had been campaigning for years on Torness, anti-dumping in the Cheviots and now Druridge from her Newcastle base, expressed relief that some action was finally taking place in the Druridge area.

Jane said, "What was frustrating me in the early 1980s was that nothing seemed to be happening. Some of my friends were busy working up an awareness of the threat of nuclear power in local trades unions. This was taking a long time to have any visible effect. It was good to see the opposition to nuclear power at Druridge centring where it should, in the area most closely affected."

Following the establishment of the Cairn, Friends of the Earth initiated other activities. Local government elections were coming up in May 1983. The first attempt to fully assess local councillors' views was made by writing to all candidates in Castle Morpeth Borough Council. Encouragingly, thirty-three of the sixty candidates written to replied and were against nuclear power at Druridge Bay. Only two, Peter Kelway and Gordon Boullous, were not opposed, and they were from the further outlying area of Ponteland.

The immediate threat to Druridge Bay was the expected arrival of a contingency of drilling rigs to do detailed test drilling in June 1983. Preliminary tests in 1979 had shown the site to be fairly suitable. These next tests were to take six months, and would be in such detail that design of the power stations could follow.

We were developing our network of advisors and supporters. Cllr Gordon Knox advised us that it might be possible to stop this unwanted test drilling by Castle Morpeth Borough Council's issuing an "Article 4 Direction" to the Secretary of State for the Environment, Tom King. Ed Metcalfe took on the job of writing to all the councillors of Castle Morpeth.

He wrote, "We are informed that the CEGB are commencing intensive test boring and environmental studies on Mr Bell's farm at Hemscott Hill, Druridge, in June. They can be prevented by Castle Morpeth Borough Council's issuing an Article 4 Direction to the Secretary of State for the Environment, Mr Tom King, in compliance with the Town and Country Planning General Development Order, 1977.

"This Direction requests the Secretary of State to ensure the CEGB has to apply for planning permission from Castle Morpeth Borough Council to do this."

Castle Morpeth Borough Council took the decision to pass the responsibility to Northumberland County Council. The County Council agreed to issue the Article 4 Direction in June.

The drillers were expected in early summer, and we presumed the CEGB would not move in until the Article 4 issue was resolved. Friends of the Earth wrote to all nine MPs between Tyne and Tweed, asking them to press the Secretary of State to issue the Article 4 Direction. Six of them agreed to do so, Alan Beith and Jack Thompson MPs in the strongest terms. The three local Conservative MPs, Neville Trotter, Geoffrey Rippon and Piers Merchant did not agree. This was much regretted by Friends of the Earth, as the request would have had much more clout with their support.

We did not succeed in getting the Article 4 Direction approved. On 21 July, Patrick Jenkin, then Secretary of State for the Environment, refused to do it. Sinking boreholes was not a serious threat, he said in his letter to the County Council, and should therefore go ahead.

This was to be expected, but something had been achieved. The extra couple of months delay for the CEGB meant that it would be difficult for them to move in and complete the six months' work before the winter weather made it impossible. An August start was simply too late. In fact, the drilling was delayed for a year, till May 1984.

Friends of the Earth were learning quickly that our efforts could produce effects. The Cairn was a success in its embryo form, and local politicians had rallied publicly to the cause. Our actions could have results, even if only to encourage people to believe there is hope, and to work for change.

One thing we learned from the Article 4 Direction was the principle of "delay". The longer we could delay the activities of the CEGB, the better our chances. This would allow us time to mobilise our forces, raise money, and help bring round public opinion. Our actions added to the work being done by dedicated campaigners nationwide, particularly those who were arguing extensively and protractedly at the Sizewell Inquiry. It all helped drive the nuclear industry into its present state of disarray.

• • •

One of Friends of the Earth Mid-Northumberland's first activities was to appoint a press officer. I took the job. Although inexperienced, I was attracted by working with the media, expecting it to be stimulating and interesting.

I blush to remember the simplicity of my approach to the job in those early days. I would write slightly different versions by hand for local papers, thinking that was what was wanted. It took hours to write the same little story six different ways.

Then I would eagerly purchase the papers to see if my offering was printed. It nearly always was, and every time it was a great thrill to see it there.

I learned that if I sent a story to several papers, it would be sure to appear in some of them. And, fortunately, I did discover that it was possible to send the same press release to all of them!

Friends of the Earth eventually moved to the sophistication of a second-hand typewriter, purchased for £20 by Bob Helling from Glaxo's rejects. To keep costs down, I would photocopy the press releases at the Community Council's office in Morpeth.

Learning to do the work of press officer was a process of trial and error, but local journalists were interested in our activities and hence helpful. I learned from them about deadlines, and developed my simple system so that if all the press releases were out by midday Monday, they would catch the weeklies in time for the following weekend, as well as the *Journal* and *Evening Chronicle* somewhat earlier.

My Special Six were the *Morpeth Herald* and the *Morpeth Gazette*, the *Journal* and *Evening Chronicle*, and the two free papers – the *Courier* and the *News Post Leader*. If a news item was particularly interesting, I would send it to papers from Berwick to Sunderland. On rarer occasions, I sent stories to the national papers too.

Gradually, in my little office in the corner of a bedroom at home, I was able to improve my technique to the point where it took an hour to write and type the press release or letter, then another hour across Morpeth photocopying it and the labels, and posting and hand delivering them. It became a morning's work.

Simple tricks like this streamlined the process. We could get press releases out every week or so, without it being too burdensome. I still had a very busy life looking after my small children, and free time was hard to come by.

Working closely with friendly journalists and copying their styles, I soon developed the knack of including the 5 Ws in the first paragraph, Who, What, When, Where and Why. Most press releases would include an appropriate quote from one of our members, and in order to do this at very short notice, I needed to know all their daytime telephone numbers. Often, for reasons of time, I would quote myself. Then I would put in the body of the story, "Bridget Gubbins, Press Officer for Mid-Northumberland Friends of the Earth, said, '............'." An advantage of providing the newspapers with a full story, including quotes, is that it reduces the chance of being misquoted.

On some occasions, journalists would ring up wanting an instant comment on what would often be a complex matter. This could be disconcerting. At first, I would say out of sheer nervousness that I would ring back in ten minutes. Then, if possible, I would discuss the response with another member of the group. It was also a useful way of avoiding the temptation to give a hasty or ill-considered response.

I also learned the use of the Letters Page. The great advantage of this is that more people are likely to read it as it goes in more editions of the newspaper. Letters often provoke replies, and this keeps the issue going.

Vince Gledhill has been involved with events surrounding the Druridge Bay story from the outset. He commented, "From the practical point of view of a working journalist, the fact that there was a single spokesperson for the group who did not mind being contacted at all sorts of odd hours was of major importance. The nature of evening newspaper deadlines means that I have often to call people early in the day. I was able to get a Druridge Bay comment before most people had started work, and that meant that the counter-argument could be included along with a pro-power station story."

On dealing with technical matters, I used myself as a guinea pig. Not having a scientific background, I realised that what I wrote had to be simple enough for me to understand it.

Vince Gledhill commented, "When technical matters needed to be explained, they were presented in language appropriate for general readers. In a perfect world the journalist should be able to take on board technical language and either translate it for the reader or verify the meaning with the writer. But in the imperfect world in which we all operate, with demanding news editors, a journalist who has five minutes to the deadline and the choice of a straightforward or complex story will usually opt for the former."

When writing press releases about fundraising events, I had to learn that it is not a matter of simply stating that they will happen. Vince said, "All the events that I can recall organised by Druridge campaigners had some news merit in their own right. No matter how worthy the cause, if I cannot make an event sound interesting to the news editor, he is not likely to believe it will be interesting to our readers."

The newspapers, local radio and TV were pleased to be in contact with us, and they made use of our easy availability. We in turn needed the publicity they gave us. We fed on each other, in the typical relationship between pressure groups and the media.

Terry Hackett, at that time a senior reporter for the *Northumberland Gazette* but currently editor of the *Morpeth Herald*, said, "Campaigners obviously realised at an early stage that public relations was going to be a key weapon in the battle, and wasted no time in establishing links with the media. The constant flow of press releases and letters has been welcome, and I am sure that it has done much to keep the issue of Druridge Bay in the public eye."

• • •

Friends of the Earth Mid-Northumberland were also required from time to time to do live radio or TV interviews. Although such work was essential for the development of our aims, none of us liked it.

Lindy Conway, as group co-ordinator, was often given the job. She was very busy with her two boys aged two and four, and instant requests for interviews could be hard to handle.

She said, "Sometimes I was given very short notice. Then I would have to arrange to have the children looked after, and get down to the studio. It was only when I got there that I had time to feel nervous. Then I'd sit and sweat. I was sure I'd sound a complete fool. When my turn came, I found the questions were quite obvious and not the difficult ones I had expected. They only ask three questions. Then suddenly it's all over, and I would wonder what it was all about."

Once a Radio 4 interviewer wanted to do a piece with Lindy. They drove to Druridge Bay to have the atmosphere of actually being on the site, and describing it to the listeners. They crouched on the dunes, trying to keep the sand out of the tape recorder, and exposed to the wind. Lindy said, "I felt I could have done it just as easily in my living room."

It became increasingly difficult to avoid doing the interviews myself. I was the person people contacted as Press Officer, and often my friends were not available at the right moment. Gradually I became used to it. Confidence came with experience.

The same problem applied to speaking in public. Shortly after Mid-Northumberland Friends of the Earth started, we were asked to speak at Kirkhill Ladies' Club in Morpeth. Not one of us would do it, and poor Don Kent had to travel up from Tyneside to cover that important engagement.

Bit by bit, several of us learned to speak at councils, students' debates, WIs, schools etc. We all went through the process of conquering our fears. For really challenging engagements, we would call on our expert advisors. It was important to match the event with a speaker at the correct level. We cultivated new speakers by taking them with those who were experienced. Their first solo talk was in a situation which was not too awesome. We found the WIs a comfortable environment for novice speakers, as a friendly welcome is always extended.

My first engagement was with Morpeth Housewives Register. The tape-slide show jammed in the middle and I did not know how to unjam it, so I improvised by chatting.

The threat to Druridge Bay put us in situations we had not planned for. Addressing Rotarians or speaking in a council chamber in opposition to an important member of the CEGB were not things any of us expected that we would have to do in our lives.

• • •

Another project set in motion by Mid-Northumberland Friends of the Earth was the Druridge Bay petition. Hilda Helling began organising a street by street collection of signatures in Morpeth. It was a good way of testing public reaction, and drawing people into conversation.

Hilda described the process in a letter she sent to the newspapers on 19 October 1984. "We got cold; we got wet; our fingers chilled; sometimes our biros iced up, and at times our doorstep discussions became protracted; but our determination to establish the truth of public feeling by a person-to-person approach brought its reward. Of those contacted, we discovered that less than 1% actually said they were in favour of a nuclear Druridge. Another 4% had no opinion. The rest were opposed, and the main reason given was concern for the environment."

Hilda also wrote many letters for the group. She was the only one with a typewriter for some time. When important public figures did not give her correspondence the immediate attention she felt it deserved, they received a sharp reminder. On 15 July, she had invited Neville Trotter to speak at the Gathering being planned at Druridge Bay in September. No reply having been received by 27 July, she wrote again. "No doubt my letter is gradually moving up the pile of mail awaiting your attention, but I wish to ask you if you will now give some priority to the matter please".

She also wrote a detailed letter requesting information to Sir Walter Marshall, chairman of the CEGB. He too was slow to reply, so Hilda wrote again. "I appreciate that you are a very busy person and no doubt did not mean the seeming discourtesy of not even acknowledging my request, but we up here are all very busy too. Can you tell me as a matter of urgency, please, how soon we can expect a reply to my letter? If you are unable to answer the questions raised, could you please tell me who can, or would you recommend us to ask our Members of Parliament to raise the matter in the House?"

Replies to both letters were quickly forthcoming.

Friends of the Earth had agreed to hold a large event at Druridge Bay. We called it a Gathering, with its quieter Northumbrian associations, rather than a rally. It was arranged for Sunday 14 September. We invited musicians, politicians from all parties, craftspeople and stall holders. Once again, Mr Storey kindly lent us the field next to the Cairn. This event required a lot of organisation from the small Friends of the Earth group. We were totally inexperienced in such matters.

The media took a great interest, and we ran a "Sing a Song for Druridge" project in our prior publicity, encouraging singers to come along and perform original compositions.

The gods were not with us that day. Long weeks of lovely summer weather changed the day before the event. Furious winds started on the Saturday, building up to gale force on Sunday with sheets of driving rain. Roads were flooded. Should we cancel or not? We were expecting Ian Breach with his crew from Tyne Tees Television, and hundreds if not thousands of people.

After a brief brainstorming session, the group agreed that we could not cancel it. Said Lindy Conway, "We couldn't have got in touch with all the artistes, craftspeople, public etc to tell them not to come. Those who struggled through deserved our response."

Ed Metcalfe said, "We decided to go ahead because we had to show the strength of our commitment. To cancel it would have been like waving a white flag to the CEGB. We couldn't let them think we were a lily-livered lot."

I stayed in Morpeth to relay messages via local radio stations to the general public, waiting tensely for a telephone call from the strugglers who were trying to put up the marquee out at Druridge Bay. If they could set it up, the event would be going on. Metro Radio announcers were waiting for the message for the 11am news. Eventually, the phone call came through from Paddy Conway. The roof of the marquee could be assembled in the open yard of the adjoining farm buildings, and the farmer had kindly given permission for us to use the empty buildings themselves. I was able to tell Metro Radio at 10.55 that the Gathering would go on. Then I dashed out to join the rest, who were straining at guy ropes and shovelling manure out of the farm buildings with old roof slates. As Ed Metcalfe commented, "At least the cow muck was old and dry!"

Hilda Helling remembers it well, "Charlie Phillips hanging on to the marquee ropes for dear life, his anorak billowing about his head and the wind threatening to lift him and the marquee clean off the ground; the thrill of seeing all those cars pouring into the National Trust car park, their windscreen wipers working at double speed; the smell of soggy wet clothes as we all squashed into the Marquee to hear the speeches and the cheerful compering of Alec Swailes".

Alastair Hardy, a potter, drove all the way from Powburn, along 20 miles of frequently flooded country lanes, to the event. He said, "My wife, daughter Lucie and I had decided to come. It was bad luck about the weather, but we weren't put off by the flooding. We were definitely against the nuclear power stations, as we had been against dumping nuclear waste in the Cheviots. We didn't expect to make great profits with the pottery sales. We helped to clear the muck out of the byres."

Carol Dixon, one of the singers, recalled, "We were almost blown away as we got out of the car, and I said to Donald 'How on earth am I meant to play the guitar in this?' The general impression which hit me as we tried to get shelter in the buildings was of the camaraderie of everyone there despite being such a mixed bunch of political persuasions, and the utter resilience of the crowd to the appalling weather. It was no surprise to me to discover that the main music group were unable to reach us due to the storms.

"I was scheduled to come on after Alan Beith. While I stood nervously at the side during Alan's speech, my kids had got themselves into the front row, especially to have a good view of Mum, and therefore really enjoyed seeing themselves on TV the following evening. My turn came at last, and Alan Beith was a difficult act to follow.

"But I don't think it mattered who was great and who was not. It was about all of us together doing our own bits in our own way for a cause that we were determined would not be lost to a faceless bureaucracy. There was a special determination which has carried on, a quiet undercurrent of ordinary people digging their heels in and saying 'No Way.'

"I realised that what mattered was that we each fought the campaign by our own methods, be they by marching, protesting, singing, political campaigning, building cairns, baking cakes, whatever. What mattered was that together our different voices were shouting the same thing, 'Hands off Druridge!'"

Although many musicians and singers had not come, Mike Tickell, one of Northumberland's most prestigious ballad singers, was there. I was very pleased when he sang my contribution to the "Sing a Song for Druridge" project, a ballad to the tune of Chevy Chase.

Then Councillor Robin Birley, Leader of Northumberland County Council, announced that the council was refusing to co-operate with the CEGB. There were cheers from the bedraggled, rain-spattered audience. Alan Beith MP for Berwick upon Tweed, Jack Thompson MP for Wansbeck, Gordon Adam MEP Northumbria and Alec Ponton from the Green Party were the main speakers.

Northumberland County Council Leader Robin Birley, signing the petition, with Friends of the Earth member Val Stephens, Tom and Sam, at the September Gathering at Druridge Bay, 1983

Later we made up our financial loss on the event by running an appeal through local newspapers, asking those who would have come if they could to send a donation. Roland Bibby, the historian, wrote in his local history column in the Morpeth Herald, "Five hundred on such a day must represent five thousand on a normal day." We took advantage of his idea by appealing for "The Funding from the Five Thousand". The response was excellent, and we recovered our losses.

• • •

Between its foundation in March 1983 and the September Gathering, Friends of the Earth had had a very busy and successful season, bringing the Druridge issue back to the forefront of public attention. Attempts by Don Kent to ensure that the group took on wide campaigning to include re-cycling, acid rain etc did not succeed. This was mainly because the threat to Druridge Bay was felt so strongly that everything else paled by comparison.

We had campaigned with little more than energy, ideas and determination. We did not have the power of unlimited money and the ear of the government, which the CEGB had. We had to match that with dedication, skill and ingenuity. For our well-paid opponents, it was a matter of their livelihoods. For us it was a matter of our environment, and involved personal expenditure which was often a problem.

We were aware that the strain on our family lives and personal resources would be hard to maintain, yet did not want the momentum we had built up to die away as had happened to the Druridge Bay Association. Clearly, a larger organisation than ours would be needed. During our work, we were making contact with many other groups and local councils, and liaison between us all was required. This became a principle aim of Mid-Northumberland Friends of the Earth, and eventually led to the formation of the Druridge Bay Campaign. This was the next phase in the opposition to nuclear power in Northumberland.

3. DRURIDGE BAY CAMPAIGN – A FEDERATION 1984

1984 was the year that the new Druridge Bay Campaign emerged, in the form of a strong federation.

Friends of the Earth's activities during 1983 had helped to bring the Druridge issue to the forefront of local affairs. The CEGB added their bit of aggravation by writing to Northumberland County Council officers and asking for detailed information, after they had already been told by the council that none would be made available. They asked for information on the sea bed, the beach and the dunes at Druridge Bay. The council refused. Then the County Planning Officer, John Lodge, received another letter asking him to pin-point views towards the proposed power station site which were considered to be most significant.

Councillor Ian Swithenbank, Chairman of Planning, said "I couldn't decide whether that was just cheek, or they couldn't read....This is getting difficult. Does Mr Lodge continue sending letters or do we reply 'leave off'? The CEGB are trying to bypass this committee by going direct to the officers."[1]

This was duly reported in the local papers. It was another annoyance to local councillors, and hardly likely to endear the CEGB to them.

Castle Morpeth Borough Council decided to follow Northumberland. They unanimously agreed that their officers must not co-operate with the Central Electricity Generating Board on any technical matters relating to a power station site.

Bad handling by the CEGB of their public relations was a characteristic from the beginning. On the other hand, it is hard to imagine them being popular, however well they handled the public, with a proposal to build nuclear power stations on a beauty spot, in a coal mining area.

Meanwhile the Sizewell Inquiry had been in session since February 1983. Opposition groups were lining up to put the case against the American Pressurized Water Reactor. Northumberland and Wansbeck councils were waiting their turn, as was the Northumberland and Newcastle Society, the local equivalent of the Council for the Protection of Rural England.

Against this background, and a miners' strike that was looking ever more likely, negotiations for a "liaison" between groups opposed to nuclear Druridge was going on. The *Courier* made headlines on 10 November 1983, BATTLE FOR DRURIDGE CONTINUES AS GROUPS UNITE. This preceded the Miners' Rally at Ashington on 19 November where Arthur Scargill was going to speak on nuclear power and the coal industry. The NUM was offering secretarial facilities to get a liaison committee going to oppose nuclear power at Druridge Bay, and

Northumberland County Council, Friends of the Earth and others would be signing up.

At the miners' rally, Hilda Helling's ditty "Stand Up for Druridge" to the tune of "Stand Up for Jesus" was played by Ellington Colliery Band, and we all sang it.

> Stand Up! Stand Up for Druridge
> Its heritage is yours.
> The C-E-G-B are waiting
> To break it in their jaws.
> Our jobs, our town, our children,
> To keep them is our goal,
> Stand Up! Stand Up for Druridge,
> Let's use our country's coal.

Arthur Scargill spoke effectively against nuclear power in general, as well as at Druridge Bay, and received a standing ovation. He was being targetted by the press as the miners' strike was expected, and I was contacted by the *Daily Telegraph* that evening to find out if I would report any outrageous statements which he had made.

Hilda's photo appeared in several papers, with Arthur Scargill signing the petition. The *Northern Echo* printed a picture of the normally straight-faced Arthur and Hilda beaming at each other in a most friendly fashion. I asked her how that came about. Hilda said, "The *Northern Echo* photographer said to me, 'Can you not stand a bit nearer to each other?' I said, 'You mean more intimately?' Arthur looked at me and I looked at him, and we burst out laughing. It was the word 'intimate'. I meant it meaning closer together, not any other way. And that was how the photographer got those spontaneous grins.

"My husband was on the train from Newcastle to London, travelling in the first class compartment on business. He walked down the passage and passed a man who was reading the *Northern Echo*. He spied over the man's shoulder a picture of me and Arthur Scargill. He said to the man, 'Do you mind if I have a look at your paper when you've finished with it? There's a picture of my wife in it.' The man said, 'Certainly not sir. You're quite welcome to it'. He was very interested and asked which was the picture. So Bob showed him, and the man said, 'Oh my God', and handed the paper to him, and wouldn't speak another word all the way to Newcastle."

The atmosphere surrounding the issues of nuclear power, the miners' strike and nuclear weapons was tense. We were trying to run a cool campaign on a local issue, keeping out of the heated wider aspects. It was a delicate job.

The NUM continued to have a great interest in the formation of the "liaison committee", and invited 48 organisations to send two representatives each to a meeting in Burt Hall in Newcastle on 31 January 1984.

Seventy one individuals, representing 34 organisations, attended the historic meeting where the DBC was founded, in the splendid Gothic hall at the NUM headquarters.

Representatives were present from nine local councils, eight trades union groups, Labour, Liberal and Ecology parties, and seven environmental groups including Friends of the Earth, the old Druridge Bay Association, Tyneside Anti-Nuclear Campaign, and the Northumberland and Newcastle Society.

The meeting was chaired by Dennis Murphy, President of Northumberland NUM. He made it clear that the purpose of the liaison was to organise and unite protest to a nuclear power station at Druridge Bay, and stick strictly to that unifying idea. The reason for this was that there were as many reasons for being opposed to the Druridge scheme as there were representatives. The NUM stated its position, but it did not dominate other viewpoints.

Northumberland NUM secretary, Sam Scott, said "We have opposed plans for a nuclear power station at Druridge Bay from the start as it would mean lost jobs in our industry, and at least a million tons of coal sterilised."[2]

The unemployment rate in the North East region was 18% in December 1983.

Following the Burt Hall meeting, there were a few informal meetings of the "liaison committee", further mailouts to a range of organisations, and finally a meeting in Northumberland County Hall, Morpeth, which established the basis on which the newly-named Druridge Bay Campaign would be run.

Two basic principles were established. Firstly, that the Druridge Bay Campaign was to be a single issue campaign against nuclear power at Druridge Bay, and secondly that it was not to be party-political. It was felt of fundamental importance that this should be a **local** campaign, which unified the opposing groups.

A steering committee was elected with representatives from each of four sections. There would be four council reps, four trades union reps, four political party reps, and four reps from environmental/miscellaneous groups. Although the numbers widened as the organisation became larger, the principle of all interested sections of the community being on the steering committee was established.

On occasions, Labour Party members or Green Party members could have taken over the political party section of the steering committee. To ensure equal representation, the groups agreed to share the seats so that Labour, Liberal/SDP, Green and Conservative groups each had a place. Everyone realised how important it was that no political party should be dominant.

Much of the strength of the developing Druridge Bay Campaign rested on decisions made in those early days. The federal structure of an association of organisations avoided the problems associated with a single group of individuals such as the early Druridge Bay Association. Friends of the Earth Mid Northumberland were already feeling the strain of the intense demands of their activities, and of maintaining an active committee from a small group. With a federal structure, when key activists move away or become worn out, organisations can put up new representatives to the committee. Consequently there has always been fresh enthusiasm appearing.

Treading the party political tightrope was obviously going to be precarious. As campaigning against nuclear power appealed far more readily, though not exclusively, to left-wing thinkers, a definite effort had to be made to include the centre ground and the right.

Gary Craig, became the Druridge Bay Campaign's first chair. He was a member of Tyneside Anti-Nuclear Campaign and of the National Union of Public Employees. He had worked hard to build up opposition to nuclear power through the trade union movement for many years. He understood the problems between various groups on the left of the political spectrum.

Gary said, "Dennis Murphy, President of Northumberland NUM, thought they should take a leading role. I agreed, because of their high standing within the trade union movement. It was also important the the Druridge Bay Campaign should be chaired by someone with links between experienced, politically aware anti-nuclear activists like Tyneside Anti-Nuclear Campaign and the trade unions' anti-nuclear position, which was newly fledged."

Gary felt as Chair he was in a position to do this. But he had to do more.

He said, "As well as contradictions between groups on the left, room had to be made for other groups uneasy about nuclear power at Druridge Bay like the National Trust, who saw it affecting their property, but had no perspective beyond that.

"I saw the task of the Druridge Bay Campaign for at least five years to build a very strong organisational platform, to create a framework within which any organisation could be involved. I had to help create political space to make it possible for Tory local authorities to support it.

"I saw myself setting conditions for the campaign to operate. Friends of the Earth Mid Northumberland had been operating on separate but parallel tracks. They had had to gain support for what could be seen to be anti-nuclear activity. The tracks met at the Burt Hall meeting. We ended up with a structure to accomodate a wide range of interests. Once the framework was ready, we created an agenda which could arrange meetings, employ workers and get into action."

Myra Blakey is a farmer's wife from the Cheviot's area, who represented the Northumberland and Newcastle Society on the DBC Steering Committee. She recounted humorously how her organisation was able to be included in the

Druridge Bay Campaign. "We saw the Druridge Bay situation as a natural continuation of the Cheviots Inquiry. As there, you couldn't even say 'nuclear power' but you're accused of being anti-nuclear. It's like waving a red rag to a bull.

"When it came to Druridge Bay, our members all said, 'Oh no, we're not anti-nuclear. We just don't want it at Druridge Bay.' We got round it by Philip Deakin saying at our AGM, 'We are against a **power** station at Druridge Bay', forget the nuclear bit."

She said, "The N & N was based at the university, started up by professional people in 1923. It's always been 'throw up your hands in horror' if you're seen to be left-wing or anti-government. But the concern for Druridge Bay is genuine. I think if it comes to the push they will venture onto the nuclear aspect of it. But in the meantime they don't really want to antagonise the membership, by appearing to be anti-nuclear."

With regard to the role of the N & N, she said, "Everyone felt that a planning application was imminent. When this happened, the N & N would have brought the Council for the Protection of Rural England behind the DBC. I think in the meantime, the N & N was quite happy that the DBC was keeping up the pressure, that they didn't have to do much."

Another of the more established organisations whose support was being sought was the National Trust. Friends of the Earth in Morpeth held a meeting with one of their representatives, and were delighted when in February 1984, the Trust's Executive Committee made a public statement of opposition to nuclear power at Druridge Bay. The Trust's statement declared that although it did not own the site of the power station, the proposal is bound to work to the detriment of its adjoining 99 acre Druridge Bay sand-dune site.

The National Trust later joined the Druridge Bay Campaign, and erected a large sign on its Druridge site proclaiming its views. The organisation has been a steadfast supporter of the Druridge Bay Campaign ever since, including offering its facilities for events.

Berwick Conservative Association finally affiliated in March 1985, and has remained loyal to the time of writing.

Structurally, the Campaign had a sound basis with its federal system. Later, another innovation improved it even more. Various people who wanted to be active in opposing nuclear power at Druridge Bay, but who did not belong to an affiliated group, asked to be members. It was decided to set up an individual membership scheme.

Dozens, scores, then hundreds of families and individuals, joined the scheme. The importance of these members is that they are highly motivated, with varying degrees of time, to help the Druridge Bay Campaign activities.

Whereas larger affiliated groups like councils paid a generous affiliation fee, and sent representatives to Steering Committee when they could, they left the routine running of the DBC to others. The individual members were of crucial importance in this respect. As the years went on, the DBC unconsciously had developed a system which was a unique blend of individual efforts and financial backing from organisations.

Individual members had to be given a form of voting rights. We encouraged them to form themselves into Support Groups in their own localities. Our network of Support Groups became organised, and was later offered places on the Steering Committee. Individual views were represented via their Support Groups.

These members became increasingly important, in lobbying their local politicians, in supporting events organised by the Steering Committee, in dispersing information, and in organising events in their own areas. They were loose groups, who seldom held meetings. Their co-ordinators tended to rally their members when specially required.

The DBC knew it had to tackle all the arguments the CEGB would promote as reasons for having nuclear power at Druridge Bay. The emerging Druridge Bay Campaign adopted the range of arguments which had already been defined by Friends of the Earth and other organisations. They were broadly five reasons. They were 1. Environment, 2. Jobs, 3 Safety, 4. No Need/Alternatives and 5. Nuclear Waste. A tape slide show which Friends of the Earth had produced went over these arguments, and further leaflets and slide shows developed these themes. We would frequently say that not all of our members are opposed to nuclear Druridge for all of these reasons, but all are for the first, the coastal environment.

By March 1984, therefore, the Druridge Bay Campaign was a seedling organisation established in the form in which it continued to grow. There were 47 local organisations paid up and affliliated, and the individual membership scheme was developing.

All this organisation and publicity were vitally necessary. The CEGB were preparing to bring their drilling rigs to Druridge Bay at an unspecified date in early summer. The Article 4 Direction had not stopped them. Public opinion was of no importance to them. Among the rest of the local population, I was awaiting their arrival with a sense of impending doom.

NOTES
1. *Morpeth Herald*, 20 10 83
2. *Morpeth Gazette*, 13 1 84

4. THE CEGB IS DETERMINED
1984

No-one knew exactly which date the drilling rigs would be arriving at Druridge Bay. We knew the CEGB needed to complete their work within the summer season, and therefore the drilling rigs would arrive in late spring or early summer, 1984. Groups associated with the emerging Druridge Bay Campaign were alerted to be ready to demonstrate against them on the Saturday following their arrival, and we were all on edge, waiting and wondering.

Early on the morning of Thursday 17 May, the telephone rang at my house. It was a journalist. The drilling rigs had arrived, in a secret, overnight move. They were there, at Druridge Bay, at that moment, ready to start drilling into **our** countryside.

Most of my friends were at work. I rang Rosie Stacy. She was at home with her little girl Victoria. We agreed that the two of us would go together with our two daughters to Druridge Bay. We had no transport, so we took the 9.25am bus to *Widdrington Inn*. From there, it is a mile and a half to Druridge Bay.

It was a cool sunny morning. Laura, my little girl, and Victoria were both four years old. It was quiet as we walked down the country lane. We didn't know what we would see when we got there. There was no sign of anything at first.

As we rounded the curve towards High Chibburn Farm, we could see people, vehicles and police markers. Then we saw the rows of enormous vehicles with the drilling rigs loaded on them behind the farmhouse. Various men were standing around, and started to approach us.

They turned out to be journalists and CEGB public relations men. Someone went to fetch George Johnston, the CEGB Stations Planning Officer. Photographers hovered round to take our pictures, perhaps hoping for a confrontation.

I had been told that the CEGB had arranged that the civil trespass law had been altered in this instance to make it a criminal act to trespass on the land where the drilling was to take place. I asked Mr Johnston about this.

He confirmed that it would be a criminal act for anyone to trespass on the land. He spoke coolly, staring at me. His colleagues were observing everything closely. I felt a sense of impotence and anger. But that eye-to-eye encounter also sowed a seed of stubborn determination in me. Whether we would lose the battle or not, it re-inforced my will to fight.

Rosie remembers, "I felt both frightened and angry. It was very threatening because the drilling machines were there, and they were very big. We were not sure what we were going to have to do. We were in the road, with such a big open space all round us. I wondered if it would ever come to lying down in front of the bulldozers."

George Johnston of the CEGB talking to Bridget Gubbins and Rosie Stacy, on the arrival of the drilling rigs at Druridge Bay, May 17 1984

She added, "Mr Johnston did answer our questions. Although polite, he was intimidating. I felt scared, not of that man, but of the power that he represented."

Photos of Rosie and me were staring out of the pages of the *Journal* next morning, along with George Johnston's, and the drilling machinery. Rosie said, "The photos were just of our two heads. I was a little bit scared of ridicule at that stage, not being sure what it would achieve. But I do believe passionately in the power of that sort of resistance. The power of an idea is enormously strong. I don't feel people need to be armed and confront physically. The fact that we went that day, and spoke to that man, and had the photos put in the paper did show on a simple level that we were prepared to do that, and bring the issues to people's attention again."

• • •

Vince Gledhill had had a tip that the drilling rigs would be coming in the week commencing Monday 14 May. He spent an eventful week before he scooped the story in the early hours of Thursday 17 May. I asked him to tell his story.

At midnight on Monday, he drove to Boldon, in County Durham, where he had heard that they might be parked at a storage depot. This turned out to be a false lead. He returned to Druridge and waited till 7.30am before telephoning his news editor to say the convoy was unlikely to be coming that day.

On Tuesday night, Vince also lay in wait. He followed a convoy of lorries from Woodhorn, which ended up going to the British Alcan Aluminium Smelter at Lynemouth. Obviously this was not the drilling convoy. Security guards at Alcan had noticed his manouevres, and alerted the police. He had trouble convincing them that he was a newspaper reporter, especially when he would not tell them what story he was after. However, they checked his car and identity with the *Evening Chronicle*, and left him. Once again there was no sign of the drilling rigs, and he left Druridge Bay at 8am.

The next night, he returned to Druridge. He said, "It seemed that the pattern of the last two nights was about to be repeated. But at about 6am there was a noticeable increase in police activity. A patrol car took up position in the car park of the *Widdrington Inn*, and others rendezvoused with it from time to time. Other police cars drove to the Druridge Bay site and parked. It was clear that something was in the wind. At 7am precisely a convoy of thirty lorries rumbled into view. They swept along the little approach road at fast speed and turned into the drilling field with a military precision.

Drilling equipment arrives at Druridge Bay, 17 5 84. *Photo Newcastle Evening Chronicle and Journal Picture Library.*

"Every lorry seemed to know in advance where it was going, and within twenty minutes they were all on site and in place. The only hitch was when one of the lorries looked as if it was going to snag a single power cable over the entrance to the field. A policeman had to climb onto the back of the lorry and hitch the cable over the projecting equipment. After telephoning my story I met George Johnston, Northern Station Planning Engineer for the CEGB, who arrived later in the morning. He told me that the speed, secrecy and timing of the operation had not been an attempt to outmanoeuvre anti-nuclear protesters, but merely intended to get a large amount of equipment onto the site with the minimum of disturbance to people in the area."

Vince Gledhill's report was in the *Evening Chronicle*, complete with photographs, the same day, and Ken Smith's in the *Journal* on Friday. The weekly papers also had full coverage at the weekend.

Councillor Ian Swithenbank, chairman of the county council planning committee, strongly attacked the move. He said they were entitled to do the test drillings, but they were still a long way from building the power station.

He said, "We will fight it every step of the way – we will leave no stone unturned. We are totally opposed to having a power station there."

He added that the area is in the middle of the miners' strike over pit closures, and here are the CEGB planning to build on top of coal. "It will have serious implications for the future of the coal industry in the north-east," he said.

Cllr Swithenbank criticised the CEGB for spending money on site details before the principle of pressurised water reactors had been decided by the Sizewell Inquiry.[1]

For the CEGB Mr Johnston defended the drilling by saying "The work is being done ahead of the Sizewell Inquiry, and before planning approval has been given because we have to work so far ahead. It takes about four years to plan a power station and eight years to build one."

Mr Johnston indicated that if the CEGB were successful with an application to build a nuclear power station at Druridge Bay, work could start about 1990 and the station would be ready by the late 1990s.[2]

The programme was moving back. In 1979 the CEGB had said that construction might be started in 1983/84 for 1990 commissioning. Extensive preparation for the Sizewell Inquiry, and its protracted proceedings owing to the number of objectors, had significantly delayed CEGB plans.

• • •

The drilling rigs were there. They were drilling in the northern half of Hemscott Hill farm, owned by Gordon Bell. For six months, they would be on site, giving us all a foretaste of our worst nightmares for the future of Druridge Bay.

The CEGB had delivered a leaflet which they called the *Druridge Brief*, via the local free paper, the *Courier*, to thousands of households in the Morpeth/Ashington/Druridge area. This four page leaflet, with maps and photos caused further annoyance.

Morpeth historian, Roland Bibby, in his weekly column in the *Herald*, called it a "Druridge Too Brief". He said, "It is notable for its short cuts, making CEGB policy sound like eternal truths, slurring outlines that should be sharp, glossing over awkward elements, and is noteworthy for its gaps and silences. What the CEGB really does generate with 100 per cent efficiency is public cynicism."

It promised jobs for more than 3000 workers, prompting the DBC to a full examination of the misleading nature of this claim. It stated the site they wanted was not in an area of outstanding natural beauty. This provoked Hilda Helling intensely, and she had a prolonged argument with George Johnston on a radio phone-in on the subject. It told us that the working life of a nuclear power station at Druridge Bay would be about 35 years, but omitted to mention what would happen to it after the end of its life.

Local papers were full of letters, news, plans and reports from local MPs, protesters and the CEGB.

Plans for a demonstration against the drilling had to be put into swift action. Groups on the mailing list of the emerging Druridge Bay Campaign had already been prepared to demonstrate as soon as the drilling rigs arrived, and they were then mailed with the date of Saturday 2 June.

Demonstrators at Druridge Bay, 2 June 1984. *Photo Margaret Eagle.*

Lindy Conway of Friends of the Earth Mid-Northumberland at the Cairn, with Jack Thompson and Alan Beith MPs, addressing people protesting about drilling at Druridge Bay, 2 June 1984. *Photo Margaret Eagle.*

MPs Alan Beith, Jack Thompson, Bob Clay of Sunderland North and Frank Cook of Stockton all attended. Hundreds of protesters gathered at the Cairn, by this time a sizeable heap of stones, from which they were addressed by the two local MPs.

Alan Beith said, "I think those who are backing the rapid development of the nuclear industry want to see it so heavily committed that there can be no turning back. There is a lesson there for us – to press our case as hard as we can in the next few months."

Jack Thompson said that after its 35 year life, he had been told by the CEGB that the nuclear power station would be decommissioned. But it would take a century, and leave a cylinder 65 metres high and 50 metres wide.

About 600 demonstrators marched from the Cairn, led by Ellington Colliery Band, along the three quarters of a mile towards High Chibburn farm. At the gateway to the drilling area we were met by security men with Alsatians and policemen. Demonstrators gathered at the farm gate, holding up placards which each spelled one letter of the slogan STAND UP FOR DRURIDGE, the theme of Hilda Helling's ditty.

This event, with photographs, was well recorded by the media. The atmosphere was peaceful, and determined. It was noticeable however that the people who came were mainly committed activists from organisations opposed to nuclear power at Druridge Bay. There was a large contingent of miners. The general public did not come. As the course of the Druridge Bay Campaign developed, this became better understood. Not everyone is comfortable with demonstrations and public parades of feeling. It can be disappointing for the organisers, but is not necessarily an indication of feeling.

Protesters objecting to the drilling at Hemscott Hill Farm, Druridge Bay, 2 June 1984. Bob Clay MP Sunderland North, and members of the Labour party, Green party, Tyneside Anti-Nuclear Campaign and NUM members. *Photo Margaret Eagle*

Carol Dixon, who had sung earlier at the Gathering, gave me her thoughts on this subject. "I'd been wondering for a while how I could get involved. Protest marches were definitely NOT my scene." Carol turned her feelings into her own way of making her point, which was singing.

We learned to accept the fact that people should use whatever talents they have to express their views. There are many ways to do it, and we continued to find ways to use all the talents at our disposal. Demonstrating is only one way to oppose an unwanted proposal. A pressure group needs people of all types, all interests, all abilities.

• • •

After the demonstration, Friends of the Earth kept weekend watch at the Cairn as the drilling went on. People who lived in the Druridge area would telephone us with details of CEGB activity. Among the stories which reached us was that the farmer who owned the land being drilled by the CEGB was interested in selling his property.

Friends of the Earth members had already made tentative approaches to Gordon Bell, the farmer, in the form of letters, but had received no reply. Messages had been passed to and fro informally via his neighbours, letting him know that we would like to discuss the issue of buying some of his land.

While in Oxford in June, I had been in touch with Friends of the Earth members who had bought Alice's Field, an untouched area threatened by road development near Oxford, and sold it in little packages worldwide. They had successfully warded off the threat to the area, and encouraged us to try to purchase the Druridge land and do likewise.

We had been making investigations into the legality and practicability of this idea, when in late July, newspapers announced that Gordon Bell had purchased Southside Farm at Warkworth, north of Druridge Bay, for £655 000. This transaction was linked to an offer from the CEGB to buy Hemscott Hill.

Ed Metcalfe who lived at Ellington, near Druridge Bay, told how he was offered the chance to buy the Bells' farm.

"Alec Bell the milkman, came round every day delivering milk. He knew my family was associated with Friends of the Earth. One day he came to my back door, probably because he didn't want to be seen. He proposed we buy the land which the CEGB wanted, and £700 000 was mentioned. It may have been a ploy to raise the stakes for selling to the CEGB. I came rushing over to see the other members, and we had a quick brainstorming session to see if we should try to raise the money."

Ed had been told by Alec Bell that we had to give them an answer in about three weeks. The CEGB were negotiating the purchase at that time.

It was school holidays, and I was camping at Edlingham on the edge of the Alnwick/Rothbury moors, with our four children. This complicated scheme had to be hatched under summer holiday conditions, and for me made harder as I was without transport, typewriter or telephone. Lindy and Paddy Conway, and Ed and Maggie Metcalfe with all their children, came to the campsite for a planning meeting. We schemed, while the children played in the stream.

After the Conways and Metcalfes left, I wrote press releases and letters by hand, and addressed envelopes for Lindy to fill, which I posted to her in Morpeth. She was obtaining legal advice from Vera Baird, our barrister friend. Paddy Conway was organising the finances, and Ed was contacting VIPs and writing letters.

I had to ring newspapers from the country telephone box in Edlingham village. Our story was interesting enough for Michael Morris of the *Guardian* to give it a top-of-the-page slot. I had to persuade him that we actually did have a realistic chance of raising £700 000, while dropping coins in to pay for the long-distance telephone call.

We sent out an appeal to all Friends of the Earth groups across England and Wales via the national headquarters' newsletter from London. We wrote to famous people who might help, including Paul and Linda McCartney, and Catherine Cookson. We contacted our own individual members, and the groups affiliated to the Druridge Bay Campaign, which had become formally set up with that name in July. (At this time Friends of the Earth were working in parallel with the newly organised Druridge Bay Campaign as one of its affiliated groups.)

It was a very ambitious task, trying to raise pledges for £700 000 in three weeks. Lindy Conway said, "We took much encouragement from success Friends of the Earth had had with purchasing Alice's Field in Oxfordshire. We couldn't let the CEGB get away with buying the land quietly, early, before a public inquiry had happened. Also we needed to show the farmer he might have a choice about to whom he sold his land. If it worked, it would have very effectively spiked the CEGB's guns, and have involved lots of people."

The local newspapers covered the story extensively over the months of August and September, relating how the funds were growing. We certainly achieved our ambition of drawing attention to the sale of the land. The three weeks passed and the land was not sold to the CEGB. Pledges were coming in slowly but surely.

Jonathon Porritt visited Druridge Bay in October, as part of Friends of the Earth's national conference which was held that year in Tynemouth. He met Alan Beith at the Cairn, duly reported by the local press. The conference took a great interest in the situation at Druridge. Activists were alert throughout the country. We were still counting the pledges at this time, and no further news had come in about the land being sold to the CEGB.

We did not achieve the goal of £700 000. By the end of September, £10 257 had been pledged. By the end of the year £12 727 had been pledged, no mean achievement for a small group, though well short of what was needed. What might we have done if we had had a couple of years!

On 13 December 1984, in another Christmas shock, the CEGB announced that they had bought 300 acres of land at Hemscott Hill farm for the nuclear power station. George Johnston told the newspapers that an application for consent to build the power station could go in by 1986, but more likely it would be about 1989. The power station would then be generating electricity by the late 1990s. Results of test drilling showed there were no major problems to prevent them going ahead. He promised that people in the Druridge area would receive a leaflet spelling out the way forward, and offering reassurance on environmental issues.

He said, "There are of course persistent voices raised in opposition, but the more information we can get to people the better."

Alan Beith MP meets Jonathon Porritt, director of Friends of the Earth, at Druridge Bay, October 1984. *Photo Ian Barkley.*

Alan Beith MP said, "It is typical of the CEGB's determination to push this scheme ahead before the Sizewell Inquiry has reported, and before there has been any clear Government decision."

The leader of Northumberland County Council, Councillor Keith Robinson, said, "The CEGB are obviously moving ahead rapidly, and I personally regard this latest move as a waste of public money."[3]

The information they had promised us arrived later that month. The pretty full-colour *Druridge Brief No 2* was delivered to thousands of households as before. It showed a tiny silver dome against a blue sky. There were no pylons in the picture. The country lane leading to the power station was narrow and unaltered. We were told that salmon would not be affected by the discharges. There was no second power station in sight. Far from being reassuring, this leaflet was yet another provocation.

• • •

Looking back over the hectic activities of 1984, I remember a particular feeling of vulnerability. We campaigners felt marginalised, even ridiculed, by pro-government and CEGB forces. It had always been tricky to maintain a sense of independence in the face of those whom it suited to cast Druridge campaigners in a radical left-wing mould. We had an awareness that we could be pictured as being in Arthur Scargill's pockets, or assumed to be doing a Greenham-Common-type action when on Cairn duty at Druridge.

But as the drilling rigs arrived, and the seriousness of the CEGB's intentions was made clear, I sensed that we were a real irritant, and being watched. When we were hatching our plans to try to buy the land, we would make and receive phone calls at numbers other than our own. Was this unreasonable? Many CND members felt their phones were being tapped at that time.

Also it was an unpleasant experience to discover that we had been sharing confidences with a man who we later discovered to be a close friend of George Johnston. He had introduced himself to me at Friends of the Earth's Tynemouth conference, and I had treated him as I would any other interested member of the public. Once he realised we knew of his CEGB connection, he boldly photographed our faces at one of our events for reasons best known to himself.

It was not a comfortable feeling to be seen as an "enemy within" while fighting to defend our own local environment. I realise now that any law-abiding citizen is potentially an enemy of powerful forces. If that person is sufficiently motivated to act as I and my friends did, he or she can be perceived as a threat, even though acting perfectly legally and democratically.

• • •

By the end of 1984, the land was drilled, and proved suitable from the CEGB's point of view for not one but two nuclear power stations. Three hundred acres of land were now in their ownership. What could struggling local people, as represented by their councils, and a federation of underfunded groups in the Druridge Bay Campaign do against the CEGB, with the full backing of Mrs Thatcher's government?

But struggles like this are unequal by their nature. Once again, the determination to keep going despite the odds won through. A conviction that we were right, if not powerful, was a strong incentive. And by their attempts to convince us that they had our interests at heart, as with the Druridge Briefs, the CEGB continued to provoke us into further action.

By this time, the Sizewell Inquiry was drawing to a close. With the inspector's report would come the expected go-ahead for the PWR programme. We needed a permanent office and to employ staff, as the struggle would be intensifying.

Money was short but ideas were plentiful. One of them was to have important repercussions in the later Chernobyl accident.

NOTES

1. *Evening Chronicle,* 17 5 84
2. *Journal* 18 5 84
3. *Journal* 14 12 84

Hemscott Hill farm, Druridge Bay. Purchased by the CEGB in December 1984. The sand dunes and sea are off the picture to the right. The nuclear power stations would fill the area of the picture immediately behind the farm. *Photo Newcastle Evening Chronicle and Journal Picture Library.*

5. RADIATION & CHERNOBYL
1984 – 1986

The Sizewell Inquiry had been running since Spring 1983. It was providing a venue for questioning the nuclear industry on a scale never seen before. It was also to decide finally on the principle of introducing the American designed Pressurised Water Reactor into Britain.

In September 1984, Dr Rosalie Bertell had given evidence for the Stop Sizewell B Association on "The Human Health Consequences of Exposure to Ionising Radiation". An American scientist and nun, Dr Bertell had been working on environmental health since 1970, and had particular expertise on the effects of ionising radiation on health. Included in her evidence criticising the CEGB approach to radiation dose limits among its employees was a point which was a great influence on the Druridge Bay Campaign.

She proposed that there should be a health monitoring system to compensate the public for harm caused by radioactive releases from Sizewell B and similar installations. She said that there was a pressing need to establish a baseline of information on health. This data should be collected **before Sizewell B is allowed to operate**, she said. Without this data, the public have no means of proving their health has been harmed, nor of making clear that a nuclear installation has been the cause. This data would be a deterrent to careless pollution of the environment, and would lay a firm legal basis for claims for compensation

Fears that radiation emissions could be associated with higher-than-normal levels of child leukaemia were by then established in the public's mind. The Black Report published in July 1984 had supported the original claim of the Yorkshire Television documentary "The Windscale Laundry" that the incidence of childhood cancer is abnormally high near the Sellafield nuclear complex.

The Black Report however had not gone so far as to indicate that Sellafield was the **cause** of the disease. Rather it concluded that discharges from the nuclear plant were not high enough to warrant the excess levels of leukaemia. Neither had BNFL accepted any responsibility for the cancers.

Druridge campaigners were not at all happy at the thought of having to live with an increased chance of local children contracting leukaemia as a result of unwanted nuclear power stations. It seemed that Dr Bertell's idea of establishing a baseline of health data in our area, combined with measuring existing levels of background radiation, was exactly what we needed. If we knew what local leukaemia levels and local radiation levels were, we would have information which we could use against the CEGB in future, should the need arise.

Among the sites proposed for further PWR development in England and Wales,

Druridge was the only site which did not already have nuclear power stations on it. We were therefore in the unique position of being able to measure existing background radiation, unaffected by a nearby nuclear installation.

We felt the information we could gather would act as a deterrent to the CEGB, as they knew that we would be able to check local radiation levels which would rise with routine emissions from Druridge nuclear power stations.

Gradually the idea took shape. In April 1985, Dr Bertell was in England for a brief visit. A few hasty phone calls were made, and it was arranged that she would visit us in Morpeth. Friends of the Earth assembled a dozen interested members and scientists in Lindy and Paddy Conway's house. Nora Cuthbert arranged to accommodate Dr Bertell overnight.

As a result of this meeting, the group agreed to pursue two aims. We would study existing background radiation in the earth, water and atmosphere around Druridge, and also assess existing health statistics regarding radiation-linked illness.

The group included Joe Cann, Professor of Geology from Newcastle University, John Urquhart, statistician and contributer to "The Windscale Laundry", Dr Nigel Connor, a biochemist, Dr Rosemary Lumb, anthropologist, and other interested members.

Many different ideas about how the required information should be gathered were discussed. In the end, the group developed a distinct preference for doing its own study as far as possible, rather than trying to get funds from the CEGB, or to attempt to set up a university project. As Dr Rosemary Lumb said, "Two factors indicate a preference for doing our own study as far as possible: one, to keep control of the data; two, our group contains the expertise necessary to design and execute a good deal of the work. Skills include radiology, geology, statistics, biology, chemistry, publicity, knowledge of the local area, interviewing and sociology."

The goals we had set ourselves were very ambitious. Rosemary agreed to co-ordinate the work, and a series of meetings followed. Setting up the project was extremely complicated, and devising a system to collate health statistics proved daunting. Rosemary commented, "There was a lack of absolutely clear knowledge, and different viewpoints emerged. There were many long meetings at which methods of measurement, data collection and interpretation were debated."

We felt penalised by our lack of funds, and by the limited time that volunteers with full-time jobs and family commitments had at their disposal.

Professor Joe Cann then gave the group a boost by writing a briefing paper which outlined how we might go about measuring radiation on a grid pattern, set in various circles round Druridge Bay.

We decided to purchase a gamma counter of the same type used by the National Radiological Protection Board (NRPB), the Mini-Instruments 6-80 environmental radiation meter. It was important to use standard, officially acceptable equipment, which would be calibrated annually by the CEGB's laboratory at Berkeley in Gloucestershire.

Rosemary wrote letters to all the trades unions affiliated to the Druridge Bay Campaign. The gamma counter would cost £599 with calibration. We were thrilled to receive a reply from AUEW-TASS (Amalgamated Union of Engineering Workers – Technical, Administrative and Supervisory Section) at Parsons, offering to buy us a radiation monitor. This donation meant that we were really established.

Other unions also donated funds. NUPE's Newcastle branch donated £100, NALGO's Tyne and Wear branch £50.

On the 5th February 1986, AUEW-TASS formally presented the monitor to the Baseline Radiation Monitoring Group, outside Parsons factory on Chillingham Road, in Heaton. When asked by a reporter why a union which has built equipment for nuclear power stations should support the Druridge Bay Campaign, Terry Rodgers, Committee Secretary, said, "We have the view that a

AUEW-TASS members (right) Terry Rodgers and Bill Caithness present a gamma radiation monitor to Rosemary Lumb, John Richardson, Bridget Gubbins and John Urquhart. 5 2 1986.

PWR is something we do not want in this area, where it would damage the coal industry and jobs. What we would require is an extension of the Blyth coal power station. We feel that this company would not stand a chance of getting turbine order for a PWR, whereas we have produced the most efficient coal-fired turbines in the country."

We knew that we had to have more than one monitor. John Urquhart came up with the idea of using our existing monitor as a source of raising funds. Therefore we wrote to our MP supporters asking if they would initiate a fund raising programme by allowing us to take a radiation reading in their garden. For this privilege, we would charge them, and later the public, £2.00. The reading in itself would have little significance. However, it would raise the profile of our work, and would have a limited use as a check against future change on that particular spot.

The MPs were very co-operative. Rosemary Lumb, John Richardson, other members and I took the monitors to the constituencies of Alan Beith in Berwick, Jack Thompson in Ashington, John McWilliam in Blaydon, Bob Clay in Sunderland and Dr Gordon Adam MEP in Whitley Bay. Wherever they went, they were followed by close media attention, with many front page stories and photographs.

As we monitored his garden, Jack Thompson MP Wansbeck said, "It is vital to have before and after figures. If we leave it till the power station is built, the opportunity is lost to make a comparison. Our successors will lack the essential information."[1]

The CEGB did not react favourably to our initiatives. In the *Morpeth Herald* in March 1986, a spokesman said that safety is the prime consideration when they are planning power stations. He claimed the CEGB take radiation measurements up to 15km from various sites at least a year before building is started, to find existing norms to give a comparison for future readings. The results are then sent to the Department of the Environment and the Ministry of Agriculture and Fisheries who can check the CEGB results.

Rosemary replied, "If the CEGB really does put such a high value on safety as their spokesman says, they have nothing to fear. Our results will be the same as theirs. However in view of the appalling environmental pollution record of the nuclear industry as a whole, the Druridge Bay Campaign does not trust official figures. We are doing our own survey as a check on the CEGB's activities."

• • •

Over the weekend commencing Friday 2nd May 1986, I had planned to take my daughters cycling round Rock Hall Youth Hostel in Northumberland. On the Tuesday before this, news of the Chernobyl nuclear accident had reached British newspapers. Worrying whether we might be exposed to the radioactive cloud, I debated whether or not to cancel our cycling weekend. However, John Urquhart informed me that the radioactive cloud was heading towards southern Europe, so I decided to go ahead with the holiday.

On Saturday, we cycled through Embleton at midday, when it started to pour with rain. We got quite wet, and went for tea and a warm up at the village hotel. That evening, at the Youth Hostel, we felt a hush go over the common room as the TV news announced that a radioactive cloud had passed over Britain, causing much fall-out in areas of high rainfall. This would have fallen on all those outdoors in Northumberland. It was a frightening thought.

This sense of fear was shared by people everywhere. Once back in the Druridge Bay Campaign office on Monday, and gleaning whatever information I could from the media, I was at the focus of local concern. The phone rang continually. Over the next few days I was nearly driven demented. I had to ring round members to help service the phone calls. Of course, we were in no position to advise the public on whether or not their children could drink milk, or eat free range eggs or home grown lettuces.

Rosemary and John's home telephone number rang continuously as well. Rosemary said, "We were the only source of information. Alan Beith asked us; and the press; nursing mothers; and even spinach growers."

It was clear that the public was in a state of panic. We did have a little information. Rosemary and John had been taking regular readings at Felton, in the field behind their house, since AUEW-TASS had given us our first gamma monitor. They were able to record the Chernobyl cloud as it passed over Northumberland. As far as we know, we were the only non-government or non-official organisation in the country who had been able to do this. We had the vital "before and after" measurements.

Rosemary comments, "It was very difficult to know how to handle these calls. We did not want to start a panic or frighten people. Yet we were running a campaign, and there really was cause for concern. We were suddenly in a very powerful position as the only people with any information."

This is what we knew. The accident had occurred on Saturday 26 April 1986. On Tuesday, the first news of the accident had been revealed in British papers. By Wednesday 30 April, the cloud had been detected over Sweden. On Thursday and Friday 1 and 2 May, our monitor had registered readings 14% higher than normal. At first, as news of the accident was still vague, we had not been sure of the significance of this reading. However, by Saturday 3 May, there could be no doubt that we had picked up a serious incident. Readings at Felton were 50% higher than normal. This gradually tapered off to 32% by the following Thursday. This much information we were able to give the general public.

We received many requests for garden readings. It appeared we had started this project at the right moment. We had just completed our publicity with MPs' gardens being read, and started doing it for the general public, so people knew we could monitor for radiation. We had continually to explain to people that there was no advantage in measuring gardens unless there had been a reading taken before the accident. It was the level of change, which represented the Chernobyl radiation, that was significant.

Nevertheless we were overwhelmed with requests. Rosemary, John and other volunteers drove miles over the region, at their own personal expense. They covered the whole area, from the Scottish border south across Tyneside, and west to Allendale, and well into County Durham. Covering such a large area, it was often only possible to do half a dozen readings in one day. They made 174 readings, the vast majority between May and June. Several people donated larger amounts than the requested £2.00, and we were eventually able to afford a second monitor.

Rosemary had a letter printed in the *Guardian*.[2] "We are apparently the only people in Northumberland who can say what background radiation was before the Ukraine disaster, and what it has been since. Our instrument is not sophisticated enough to distinguish between the nuclides contributing to current increases, and we certainly cannot tell people whether or not is it is safe to drink milk." She concluded, " The only solution for those who are sceptical of official facts and figures is to undertake their own research."

This led to enquiries from groups all over the country. They all had to be answered. Some were seeking technical advice which we were not qualified to give. The letter also led to contact with two computer engineers from Tyneside. These meetings helped initiate the Argus project, a computer-linked constant radiation monitoring system, which is now established.

Our figures broadly agreed with those released by the National Radiological Protection Board on Wednesday 7 May. In the *Guardian*, the NRPB was reported as saying that the radioactive cloud affected southern counties on Friday and Saturday 2 and 3 May, then moved slowly north. Peak measurements of radioactivity in the air were between one and ten becquerels (the standard unit of radioactivity) per cubic metre on Friday and Saturday. By midday on Sunday levels had fallen to one-tenth of the peak.

Already planned for June that year was a national conference on the effects of low-level radiation in Gloucester. It was organised by SCAR, Severnside Campaign against Radiation, who were concerned at the increasing number of leukaemia cases in children near the series of nuclear power stations on the Severn Estuary. Experts from across the world gave lectures and workshops on radiation and health. Four members of our group attended, and we met other people who were, like ourselves, initiating radiation and health monitoring projects across the country.

Our group slowly developed, learning all the time, gaining understanding about radiation and how to use the monitors. The problems of finding time and energy to do the work were continuous. A group of scientists such as ours with very strong and differing ideas can be difficult to co-ordinate, and there were many uncertainties and differences of opinion. Nevertheless, the monitoring of gardens and regular readings at the Felton site still was going on.

In October Blyth Valley Borough Council commissioned our group to take a series of readings for them over the winter of 1986-87. After negotiation, the council agreed to donate another 6-80 monitor to us for this service, and £500 for expenses. The work was mainly done on consecutive Saturdays, for several weeks.

Meanwhile, a London scientist, Dr Charles Wakstein, agreed to prepare a report for us on the consequences of a full-scale meltdown at a future PWR nuclear power station at Druridge Bay.

His report was given to us in October 1986. His findings shocked everyone. Although Druridge Bay may seem to be fairly remote, in fact several towns with large populations are within ten miles of the site. The Tyneside conurbation also lies within a 25 mile radius. Therefore the CEGB's choice of Druridge Bay as being an isolated site was completely inappropriate.

Using officially-published figures, Dr Wakstein calculated that if there should be a meltdown at a Druridge Bay nuclear power station and the wind was blowing in the direction of Morpeth and Ashington, the following deaths would occur: if the weather was wet, 100% of the populations of Morpeth and Ashington would die, unless evacuated for 68 or 77 years. Figures were published for all local towns. In the case of Newcastle, 29.2% of the population would die unless evacuated for 44 years, Alnwick 74.1% unless evacuated for 62 years, Berwick 6.8% unless evacuated for 26 years.

The type of accident referred to is called a UK1 accident. This is extremely severe, and typical of PWR nuclear power stations. It is worse than the accident at Chernobyl, but not the worst possible.

At the time of publication of the Wakstein report, the CEGB responded by saying that Dr Wakstein's figures were "unnecessary alarmist claims." Long before Chernobyl, the CEGB said, it had been decided that the RMBK type of nuclear power station such as the one at Chernobyl would not be built in the UK. Comparison between the Russian plant and what was envisaged for Druridge was therefore irrelevant.

The spokesman continued, "In the worst possible accident which we say could occur at PWRs such as Druridge, we would evacuate in a radius up to one and a half kms from the site. This is a low possibility accident."

As the continuing problems in the USSR reveals, the effects of nuclear accidents on populations and health are catastrophic. Dr Wakstein calculated that crops would be banned for 100 years in areas up to 25 miles from Druridge, milk banned for 50 years in areas up to 16 miles from Druridge, and meat banned for 50 years from areas 10 miles from Druridge. These calculations now seem almost optimistic, considering the contaminated areas hundreds of miles from Chernobyl in Byelorussia, Ukraine and the Russian Republic. Sheep grazing on hills 1500 miles from Chernobyl in Wales and Cumbria are still unable to be sold in 1991.

His evacuation figures also now seem perfectly credible. In the USSR 135 000 people have been evacuated from areas around Chernobyl. (Soviet IAEA Report, August 1986) As a result of the accident, 600 000 people have been classified as having been "significantly exposed" to radiation, put on a special register of people whose health will be monitored for the rest of their lives. Thousands of people outside the specific 30km (18 mile) exclusion zone round the reactor need to be evacuated. These people continue to absorb an increasing dose of radiation through eating contaminated food and breathing contaminated air.

The Chernobyl accident and Dr Wakstein's report increased the level of opposition to nuclear power at Druridge Bay. The public had swung more than ever in our direction, and the potential importance of the Baseline Radiation Monitoring Group was increasing.

During 1987, following the hectic activities during and immediately after the Chernobyl accident, the group gradually became tired. Rosemary commented, "I had worked as co-ordinator for two years. Nearly two hundred garden readings had been taken, a very time consuming and repetitive procedure. The original volunteers seemed to be falling off. The group had reached the stage of owning three monitors. It seemed a good time to draw a line and let others decide how to set up regular monitoring with new committed volunteers, new energy and new directions."

Over years of campaigning we had grown used to the ebb and flow of people's energies. It was always a sad moment when committed members like Rosemary decided she could not go on. We cast around for a new co-ordinator.

Helen Steen, a geologist who had worked for local government for several years, took on the job. She said, "I had a little girl, Iona. She was 18 months old. I wanted to be involved in the Druridge Bay Campaign, as I had some spare time on my hands. My professional training and scientific background could be useful."

Libby Paxton, psychologist, who had taken on co-ordination of Morpeth Support Group, intensified her activities after the Chernobyl accident. Her interest was in the health aspects of radiation. She recalled, "The Chernobyl crisis occurred in the very week that I was weaning my daughter Sophie. Normally I would have introduced cow's milk, but was dissuaded by my GP and friends. I realised that eventually both my children could be ingesting contaminated foods. I felt so powerless. One cannot starve a child in case the food and drink is contaminated.

"However, I also felt angry enough to look into the problem more. I went with Toni Stephenson, another DBC member, to see Dr Alan Craft and Stan Openshaw at the Child Oncology Unit at the RVI to ask about the relationship between leukaemia and cancers, and radiation levels. Later I went to Stirling to the Third International Standing Conference on Low Level Radiation and Health, in June 1988. At that conference, there was the first public mention of leukaemia in children being related to radiation exposure of fathers. I gradually became aware of the seriousness of the problem."

1986 was the year when public consciousness of the dangers of nuclear power was at its height. As well as the Chernobyl accident, there had been a series of leaks at Sellafield during January, February and March, followed by a damning report by the Commons Select Committee on the environment. Sellafield had become an international scandal. There were leaks at Dounreay reprocessing plant in Caithness in May, and faults at Dungeness and Hinkley nuclear plants in May and June.

Anxiously, the nuclear industry tried to regain the favour of public opinion. Its immediate reaction to the Chernobyl accident was to claim that "it couldn't happen here".

The Sizewell Inquiry report, due later in 1986, was bound by its terms of reference to ignore anything that happened after the Inquiry had finished, including Chernobyl.

The nuclear industry brazened out the consequences of the accident, relying on the short memory of the public for drama, however intense. The industry is enduring and patient, and has plenty of time to sit out problems. The focus of attention has moved from the safety aspects of nuclear power, so prominent in 1986, to its economics in the late 1980s and early 1990s.

Nevertheless, since 1986, the Druridge Bay Campaign's Radiation Monitoring Group has persevered.

John Griffiths, a member from Glanton in Northumberland, rallied up members from Whittingham Vale and Alnwick Support Groups. John is a materials scientist, well qualified to coach members on radiation, and to select suitable sites for monitoring. Members took readings on four sites, three times a week for ten minutes, and one site for an hourly reading every six weeks. The group has an extensive data base now, and continues regularly.

John said, "I never felt our main purpose was as an early warning system for another Chernobyl. We couldn't afford to set up that sort of operation. We were basically getting background readings at specific sites in the Whittingham-

Alnwick-Amble areas, to learn over the long term what was happening. The more years we take the readings, the more reliable the figures are."

Fiona Hall is one of the team of people monitoring around Whittingham. She said, "Each volunteer only goes out once every two months or so. People like to do it because it is something practical. When you actually measure radiation, you feel you are in contact with the reason for the existence of the Druridge Bay Campaign. We also had to learn a bit about radiation. Over the months we saw a pattern in the readings. They varied considerably across the different sites, but on each site the reading kept within a very small range. We could see we would easily notice if something dramatic happened."

New members Andrew Home and Niall Urquhart took over the co-ordination of the monitoring group from the central area round Morpeth. They have four sites which they monitor weekly, at the National Trust Car Park at Druridge Bay, at Plessey Woods Country Park, in Lancaster Park in Morpeth and at Ashington High School.

Andrew said, "It was quite a big commitment of our time. However we have collected a good data base. I collated both our readings and Alnwick's on computer. The next set of monitoring sites needs to be firmly established in the Tyneside area, and is a forthcoming project."

We have lost Andrew, as he moved to his new job in Wales. He has been replaced by John Duggan, a physics teacher. Pete Sendall is working with Niall Urquhart on Tyneside to develop readings for that area. The work goes on, thanks to willing volunteers.

Although the original aims of the Baseline Radiation Monitoring Group proved overambitious on its health aspect, the radiation monitoring programme has been a success. There are up to twenty Druridge Bay Campaign members who can handle the equipment, who could take readings in case of an emergency, who were on alert to do so if required during the Gulf War, and who are assembling information in case of a future nuclear power station at Druridge Bay.

These dedicated volunteers are quietly working away in the background. If there is another nuclear calamity, or should Nuclear Electric put in a planning application for Druridge Bay, the public will appreciate their work. Their knowledge will be as much in demand as it was at the height of the Chernobyl accident.

NOTES

1. *News Post Leader*, 28 2 86
2. *Guardian*, 10 5 86

6. POLITICAL DEVELOPMENT
1985 – 1987

The Druridge Bay Campaign was never party political in its approach. Members came from all sections of society. But the Campaign was very political in its activities. Gaining influence was very important. Seeking the support of those in power, who could help make decisions which would say yes or no to nuclear power at Druridge Bay, was always a priority. This had to be done both locally and in the national arena. Having support from local authorities was also a potential and vital source of funds.

On the one hand, we sought help from local authorities who could support the Druridge Bay Campaign. On the other hand we were helping them by concentrating the resources of volunteers to achieve the same ultimate objective: keeping the nuclear industry from Druridge Bay. If the councils gave us financial contributions, we could be a mutually effective partnership. As a pressure group, the Druridge Bay Campaign was much more targetted at the government in far-off London and the monolithic, national CEGB than at local councillors or MPs. It was a local cause, with our opponents clearly seen as outsiders.

In 1984, the CEGB had announced that the Druridge site was suitable for nuclear power stations, and they had purchased 300 acres of Mr Bell's farm. They had further infuriated local people by buying properties at Druridge Bay to the value of £500 000 in September 1985. Unfortunate home owners who lived in the charming modernised stone cottages at Druridge Bay had found they might be living literally next door to nuclear power stations. When they tried to sell their houses, no one would buy them. They were very relieved when the CEGB took them off their hands, but it was seen by most people as the CEGB purchasing properties for their own purposes, and buying out the opposition.

County planning officer John Lodge said, "It is Big Brother tactics. They have got the money so they are going to buy out those properties as they have already bought out the farm or the site of the power station. When you have not even got an agreement in principle to go ahead with a nuclear power station, then to spend money on this does undermine the whole system of the planning process."[1]

The seriousness of the CEGB's intentions at Druridge Bay could not be doubted. The Sizewell Inquiry was drawing to its close. It ended in Spring 1985, after 26 months, and 340 days of taking evidence. The Inspector's report was expected to recommend the building of Sizewell B, the start of the British PWR programme. Against this background, the new Druridge Bay Campaign began to develop local political support.

The DBC was holding monthly meetings. It clearly needed an office base from which to conduct operations on a larger scale, and workers whose energies could be spent concentrating on its objectives. Funding for the office was where the support of local authorities was to be of crucial importance.

During 1984, Friends of the Earth in Morpeth had subsidised a small office in a spare bedroom in my family home, for which they paid £15.00 a week rent. My four children were growing up, with the youngest, 5 year old Laura, coming home for dinner every day at noon. The threat to Druridge Bay had become an important part of my life, and I had not started looking for work outside the home. I was able to spend two hours each morning working in the office. As well as the rent for the office, I received a stipend of £10 weekly from the Druridge Bay Campaign to send out regular press releases. These small remunerations helped me to work for the Campaign, and the little office served as a base for co-ordinated activities. However I was keeping my eye out for a space which was more public and well equipped.

I was taking all our photocopying to the Community Council for Northumberland's office at Sanderson House in Morpeth. They allowed voluntary groups to use the copier at cost price. The Community Council was soon to be moving to larger premises, and I found out that some office space would be available for rent to outside organisations.

Gary Craig, our Chairman, was keen to employ workers, and together with Clare Barkley, our treasurer, set about trying to obtain funding for office and worker salaries. They produced a glossy booklet, setting out the aims and function of the Druridge Bay Campaign. They set a great deal of importance on excellent presentation. The booklet was aimed at local authorities who were already affiliated to the Campaign, to encourage them to give a large annual donation towards our funds.

Local government operates its finances on an annual cycle. Both Gary and Clare were familiar with this, and the committee system. Gradually we obtained agreement from our supporting councils for varying sums of money. By the end of 1985, Gary and Clare were confident that enough money would be available to employ workers. They advertised for a Publicity and Information Officer. We began negotiations for an office in Tower Buildings, the new home of the Community Council for Northumberland. With luck, the Druridge Bay Campaign could have its own base and workers in central Morpeth by Spring 1986.

There were over 20 applications for the position, and the candidates were of the highest calibre. I applied. To work for the Campaign was my greatest hope and desire at that time. However, there was no question of a straightforward appointment for me. The job was advertised, and shortlisted. I had to go through the procedure and be interviewed like all the other applicants. I was extremely relieved when Gary told me that the job was mine if I wanted it.

Consequently, in April 1986 I was able to start work in the freshly-decorated Druridge Bay Campaign office. Its location in Morpeth town centre was conveniently near my home, and made attendance at evening meetings and general access for members very simple.

At the Annual General Meeting in September that year, funding for office and workers amounted to £9800.00. It had come as follows:

Blyth Valley Borough Council	£500
Gateshead MBC	£500
Newcastle City Council	£500
North Tyneside MBC	£2500
Northumberland County Council	£5000
Sunderland Metropolitan Borough Council	£300
Tyne and Wear County Council	£500

In September 1986, a second post was advertised. Wendy Scott was appointed as Education and Development Officer. She lived in Hauxley, a hamlet at the northern end of Druridge Bay.

By the time of the 1987 AGM, new councils had been added: Wansbeck District Council, who donated £3000 for two years; Castle Morpeth Borough Council who gave £500; and and South Tyneside £500. Most of these councils continued to give these donations on an annual basis, after our original application. Sunderland MBC dropped off, as their donation had been a single contribution, and Tyne and Wear County Council was abolished. However, the ongoing funding gave us a reasonable degree of security, enabling us to plan ahead.

The councils had many reasons for agreeing to donate money. All their different political structures affected their decisions. In 1986, the largest political group on Northumberland County Council was Labour, but it did not have an overall majority. Liberal and Conservative groups had to agree about the usefulness of this funding. An application to oppose nuclear power nationally rather than nuclear power at Druridge Bay would have caused difficulties.

Cllr Bill Ashbridge, (Lab), Chairman of the Planning and Economic Development Committee, took a great interest in the Campaign, and helped us achieve the £5000 grant from the County Council. He said, "I was grateful that someone showed an effective interest in the preservation of our coastline. Any power station would have been an eyesore to the area. However, if funding had been used for any other field than specifically Druridge Bay, we would have been unable to budget for it."

At the County Council meeting in February 1986, the Liberal/SDP group was supportive. Cllrs Anthony Castle and Ian Hinson argued that the money would support the much-valued DBC radiation monitoring programme.

The Conservative group opposed nuclear power at Druridge Bay on planning grounds only. They were concerned about the choice of the specific site at Druridge Bay. They were not against nuclear power as such, and not all the councillors wanted money to go to the Druridge Bay Campaign.

Cllr Derwent Gibson (Con) felt it was "money mis-spent". If an application should be made to develop a nuclear power station at Druridge Bay, the council had the machinery to object, he had said. "I move we try to recoup it," he said, at the council meeting.

Council leader Keith Robinson disagreed that the £5000 was mis-spent. The Druridge Bay Campaign organisers were carrying out study work that the council had not been able to do. This included work on radiation levels. "The CEGB has spent thousands and thousands on publicity broadsheets to promote the idea of a nuclear power station, for which we have to pay in our electricity bills. This is a small contribution to offset the propaganda," he said.

Cllr Gibson's proposal to recoup the money was rejected by 36 votes to 13.[2]

Castle Morpeth Borough Council, the planning authority for Druridge Bay, took longer to commit itself fully to funding the DBC.

Liberal Democrat Cllr Janet Cann became Castle Morpeth's representative on the DBC Steering Committee. She explained the process. "Although there was a strong desire among the councillors to preserve Druridge Bay, at least two thirds of them were not against nuclear power as such. There was an initial feeling that the Druridge Bay Campaign was more anti-nuclear than anti-power station of any kind, and a view that the DBC was too political with a big Labour 'P' for a non-political council to give financial support."

She was able to emphasise to the DBC steering committee that Castle Morpeth's environmental stance of "no industrial development of any kind" would, if backed by the DBC, result in funding and support by her council. It was a similar argument to that which had taken place with the Northumberland and Newcastle Society earlier. As vice-chairman of the Planning Committee, she was then able to reassure her fellow councillors that her views were understood and accepted by the DBC.

She said, "It was also recognised increasingly by all that the DBC was not a wild organisation but doing excellent work in organising its material to support the case cogently against the CEGB. Trevor Hulbert (Liberal Democrat) backed by me proposed that we should commit ourselves positively to the DBC by giving it £500 annually, a huge sum for a council of our size and caution, instead of the usual £50 affiliation fee. I was delighted that David Adams, the Tory chairman of the committee backed the proposal whole-heartedly from the chair and was supported by the Tories. The result was a unanimous acceptance, a moving moment for those of us who had worked so hard to bring it about."

"There is a great significance in that unanimous vote and the benefit such all-party support of the planning authority must give to the cause. It is not perhaps a co-incidence that other non-Labour councils in the area then followed suit."

Cllr Cann was referring to the later affiliation to the Druridge Bay Campaign of Tynedale District Council and Berwick upon Tweed Borough Council, which followed in the next year or two.

Wansbeck District Council is an almost entirely Labour council, and it was within their remit to support the Druridge Bay Campaign. Cllr George Ferrigon told why he supported our application for funds. "Although Druridge Bay is not in our district, nuclear power there would have an effect on the mining industry. It was our duty to look after people still working in the mines. Once we realised the full effects of nuclear power, other concerns rose to the fore. We discovered the consequences of a nuclear accident on our area. We learned how the power station would affect the actual environment at Druridge Bay, which is used by our community."

When asked what was the value to Wansbeck District Council of the Druridge Bay Campaign, Cllr Ferrigon said, "The council fights at a hundred fronts. The DBC concentrates on this one issue. It is not party political, and it is fighting our cause on job losses in the mining industry."

He adds, "Pressure groups have got some clout. The Druridge Bay Campaign is prominent in our area. For five years I have watched the DBC grow up. Maggie Thatcher will have had a little book somewhere with DBC on, with a little red ring round."

Cllr Ed Derrick is Chairman of the Nuclear Issues Working Group on Newcastle City Council, now incorporated in the Public and Environmental Protection Committee. Newcastle City Council is also very strongly Labour.

He said, "The City Council's anti-nuclear policy meant we supported those who campaign against nuclear power in general. We were concerned about Druridge Bay because of effect on local employment and the regional economy, and the long term effects of nuclear power, specifically locally. There was also the problem of the premature closing of Blyth power station, and the damage to mining and local engineering. Parsons needs the support of the coal industry.

"Newcastle City Council's policy objectives were served by supporting the Druridge Bay Campaign. Sometimes, specific activities can be seen as political campaigning, and we can't take part. Voluntary organisations can take a more up-front stand, they concentrate on the single issue. A specialist organisation is required."

Newcastle City Council also donated worker time. Adrian Smith, one of their planning officers, was allowed to include professional support for the Druridge Bay Campaign as part of his workload.

The successful gaining of funding from our local authorities was of fundamental importance in the next stages of our campaign. It is quite clear that this would not have been forthcoming if the councillors had not been expressing the direct wishes of their electors. Since 1979, when the CEGB first announced their interest in Druridge Bay, it had been increasingly clear that the vast majority of people in the north-east did not want nuclear power in their area. The councillors were confidently representing the population by supporting the Druridge Bay Campaign.

Other councils became affiliated over the years. This had required a good deal of careful co-ordination work involving meetings and sometimes lobbying.

By December 1989, fourteen district, borough, metropolitan and county councils were affiliated. They were spread from County Durham in the south over the urban area of Tyneside, north and west across Northumberland and up to the border of Scotland. The original strong reaction against nuclear power at Druridge Bay had been disjointed and unorganised. The Druridge Bay Campaigners had drawn together a united and formidable organisation.

• • •

Councils affiliated to the Druridge Bay Campaign (September 1988)

Political work in the national arena was developing concurrently with the co-ordination of local support. During 1987, there were two particularly good opportunities to focus public attention on the threat to Druridge Bay, the publication and debate on the Sizewell Inquiry report early in the year and a general election in June.

The long-awaited Sizewell Report went to Peter Walker, Secretary of State for Energy, just before Christmas. It was released to the public on 26 January 1987. The Inquiry Inspector Sir Frank Layfield gave a clear recommendation that the Sizewell B PWR should be constructed in the national interest, despite his many criticisms and qualifications. "The CEGB has discharged the onus of proof that national need overrides the local interest in favour of conservation," he said. "In my judgment, the expected national economic benefits are sufficient to justify the risks that would be incurred."

He criticised the CEGB and the Nuclear Installations Inspectorate on detailed safety issues, but nevertheless concluded that such matters could be left in their hands.

He was scathing about the CEGB's economic case, and produced his own calculations, concluding that Sizewell had a one-in-four chance of losing the nation money.

As the report was unable to draw conclusions from any information made available after the inquiry closed in spring 1985, it ignored the 1986 Chernobyl accident, and asserted that an uncontrolled release of radioactivity was only likely to happen once in a million years.

Equally it ignored the slump in coal and oil prices, which by 1987 suggested that building a coal-fired power station would be cheaper.

The report was not a surprise. It simply meant that the battle had to go on as we had anticipated. The Druridge Bay Campaign had one of the best speakers in the British environmental movement to comment on the Sizewell Report, and its impact on Druridge Bay. Jonathon Porritt had agreed to speak to a public meeting on 29 January, in Newcastle Civic Centre, on our behalf.

Jonathon emphasised the out-of-date nature of the report. Referring to Chernobyl, he ironically quoted Sir Frank's conclusion that "a catastrophic accident would almost certainly never occur".

He said, "Layfield tells the CEGB that they should not proceed with the construction of a reactor unless the safety limits are below the level of tolerable risk to ordinary people. Yet opinion polls tell us that between 60 and 70% of people in this country are opposed to an increase in nuclear power."

On Layfield's statement that Sizewell was acceptable if the economic benefits justified the risks, Jonathon pointed out that projected costs of a barrel of oil and a tonne of coal in the year 2000, as used in the inquiry, were already out of date, and were at least half as low. "This completely undermines the economic argument," he said.

Publication of the report was to be followed by a debate and vote in the House of Commons. Friends of the Earth in London were encouraging their members to join in a national lobby of MPs in February. Druridge Bay Campaigners planned to join in this lobby, with their own theme. Targetted at North East MPs, the motto was, "A Vote for Sizewell is a Vote for Druridge". The aim was to persuade all North East MPs to vote against Sizewell B.

Plans for the lobby had been prepared. Members and affiliated groups had been encouraged to write to their MPs, and sign up for the lobby. The Northumberland and Newcastle Society wrote to 21 North East MPs, asking them to register a NO vote, as "we believe that a decision to go ahead at Sizewell will only bring Northumberland nearer to suffering a similar fate".

TASS-Joint Office Committee of NEI Parsons wrote to all North East MPs, asking them to adopt a policy of opposition to the PWR as not in the interests of the North East. Hexham Town Council wrote to Geoffrey Rippon to ask him to vote against the PWR. North Tyneside Council wrote to both Neville Trotter and Ted Garrett, MPs for their area, pointing out safety risks, and asking them to vote against the Sizewell PWR.

MPs Alan Beith, Berwick upon Tweed, and Jack Thompson, Wansbeck, as usual helped and encouraged us on this venture. Alan Beith, as constituency MP for the Druridge Bay area, booked committee room W1 in the House of Commons ready for our visit.

We had written to all local MPs, asking for their agreement to vote NO in the debate, and received many replies. We made big posters from the replies received, putting MPs in three categories. Under one heading saying "THESE MPs WILL VOTE NO TO SIZEWELL" went the names, big and bold, of those who clearly stated their opposition. Another big poster said, "VOTE NO TO SIZEWELL. THESE MPs HAVE NOT SAID NO." Under that heading went those MPs who we expected to vote YES. And a third said, "VOTE NO TO SIZEWELL. THESE MPs DON'T KNOW?" In the last category we listed MPs who had not replied to our letters, or who had not made their position clear.

Jane Gifford made twenty big copies of our DRURIDGE NO! emblem. They were mounted on bamboo sticks. Together with our banner and the posters, we had a good display.

A bus load of campaigners from the North East travelled to London. Most of us had never been on a lobby at the House of Commons before, and were not sure what to expect.

We arrived at 1.50 pm outside the Houses of Parliament. The crowds were milling, tourists gazing, and campaigners against nuclear power from across the country were lined up waiting to see their MPs. We tumbled out onto the pavement, with all our posters and banners, and were relieved to see our MPs Alan Beith and Jack Thompson waiting for us.

TV cameras and reporters were waiting, and filmed and interviewed us. We had such a good visual display that we attracted a good deal of attention.

We were then whisked past the other queuing lobbyists, and allowed by the policemen at the entrance to store our posters and banners in their tiny cloakroom.

We were led through the ancient halls to our Committee Room, where we gave a prepared presentation of our case to Labour MPs from across the North East. Alan Beith chaired the meeting. Unfortunately, no Conservative MPs attended.

Four more North East Labour MPs agreed to back the Druridge Bay Campaign. Nick Brown Newcastle East, Giles Radice Durham North, Ernest Armstrong North West Durham and Ted Garrett Wallsend agreed that they would vote against Sizewell.

We were particularly pleased that Ted Garrett had joined the ranks of supporters. In a letter dated 2 January the same year, to Gordon Steel, he had said, "It would be premature to oppose totally nuclear for energy and this would include the Druridge Bay scheme." He had based his argument on job losses.

Cllr Wendy Sayer, of Castle Morpeth Borough Council, who attended the lobby, said she had become a councillor with the sole intention of fighting for Druridge Bay. She said, "I was on the council's planning committee, and was asked to go to the lobby on their behalf. I was pleased by the support and turnout of Liberal and Labour MPs, and surprised no Tories turned up for the meeting. But I felt we were batting our heads against a brick wall, because the government's policy is for more nuclear power. I felt we were doomed, even though up in the North I don't know any councillor in favour."

After that, some campaigners met their MPs informally or by appointment, while others went outside to participate in the "Photo Call", which had been organised by Friends of the Earth.

Photographers and television reporters pounced on us. Liberal leader David Steel and shadow spokesman for energy Stan Orme posed for photographs with our group. My six-year-old daughter Laura remembers Stan Orme giving her a packet of Polos. It was the high point of her day. She was photographed with David Steel's arm round her, while she sucked her sweets.

Cllr George Ferrigon, who attended for Wansbeck District Council, recalled his impression of the event. "I was quite surprised at the reception we got. I expected no-one to take much notice of us, but we were well received. It made an impact on the MPs. Even the Tory MPs, who didn't come to our presentation, knew we were there. And being on TV has an impact on people at home, which has an impact on MPs. We had to change public opinion, and that is happening."

One rather unfortunate result of our listing of names of MPs was that some of them felt we had wrongly described them. Some of the MPs who had not replied to our letters did not like to see their names under "don't know", and had been offended. Jack Thompson confided this to me, and also explained that the tone of one of our letters, requesting them to come to our meeting, was rather uppity. I was sorry that we had made these errors. As a consequence, though, more North East MPs did reply to us afterwards, and we were able to add nearly all of their names to our list of supporters.

In the informal meetings in the House of Commons, some interesting contacts were made. Gordon Steel, our treasurer, met Lord Glenamara in the Strangers' Bar. He was better known locally as Ted Short, and had been MP for Newcastle Central between 1951 and 1976. He had held several top posts in Labour governments, as well as being deputy leader of the Labour Party. He agreed that he would do anything he could to support our campaign.

Piers Merchant, Conservative MP for Newcastle Central, met some of his constituents from our group. He told them that he supported the development of Sizewell B. But when pressed, he said, "If 10,000 of my constituents beat a path to my door, I would vote according to their wishes."

Following the "Photo Call", Friends of the Earth London had arranged a meeting of campaigners. For the first time we exchanged news with people fighting against PWRs in the other areas.

In the Grand Committee Room, reserved by London Friends of the Earth for the occasion, there were more informal meetings, followed by a two-hour debate in the evening. Prominent speakers included Malcolm Bruce, Alliance Energy Spokesman, Robin Cook MP, and Frank Cook MP. They spoke against the representatives of the nuclear authorities and pro-nuclear MPs.

The bus party left for home at 9.00 pm (after a 6.00 am start), arriving in Newcastle at 3.30 am next day.

For campaigners like ourselves, such an exercise was largely about being seen and heard. Although we could not change the government's mind, we can let the powers that be know we were determined to oppose their plans. It was our first, but not our last visit, to Parliament. We were learning the arts of successful campaigning.

• • •

Prior to the expected date of the Sizewell debate in the House of Commons which would be following the lobby, the Druridge Bay Campaign's Newcastle Support Group joined forces with Tyneside Friends of the Earth and Greenpeace groups. They decided to take Piers Merchant MP up on his challenge that, if 10,000 of his constituents beat a path to his door, he would vote according to their wishes. They visited his constituents in Newcastle Central.

Alan Batty, then co-ordinator of Friends of the Earth said, "In the up-market area round Osborne Road and Jesmond Dene Road, at almost every single door I knocked on, people signed our request that Piers Merchant should vote 'No' to Sizewell without hesitation. In the student bed-sit area round Cavendish Place and Manor House Road, about 90% signed."

This made some useful publicity from our point of view. TV picked up the story, showing Piers Merchant paddling in the waters of Druridge Bay with his family, declaring how industrial architecture would look nice there, and how favourable it would be for local jobs.

He was one of the few local politicians in the North East who publicly declared that he was in favour of nuclear power at Druridge Bay. In the general election which followed a few weeks later in June, he lost his seat.

• • •

After the lobby came the debates in the House of Commons. On 11 May 1987, North East MPs had their chance to state their views on the Sizewell reactor, with its consequences for the British PWR programme, and Druridge Bay.

Jack Thompson MP was the only local MP who was able to speak in the debate itself. He stated categorically that he was a sponsored miners' MP, and a patron of the Druridge Bay Campaign. He emphasised the poor economics of nuclear power, the nuclear waste problem, and the fact that nuclear power workers could bring the nation to a standstill overnight.

In the vote that followed, 15 North East MPs voted against Sizewell. They were all Labour, with the notable exception of Alan Beith. None of the three Conservative MPs had been able to bring themselves to vote NO, to our great disappointment.

Voting among local MPs was as follows:

Against Sizewell:
Gordon Bagier
Alan Beith
Roland Boyes
Nick Brown
Robert Brown
David Clark
David Clelland
Bernard Conlan
Donald Dixon
Jack Dormand
Derek Foster
Ted Leadbitter
John McWilliam
Giles Radice
Jack Thompson

For Sizewell:
Piers Merchant
Geoffrey Rippon
Neville Trotter

I sent a press release out on the verdict, referring to the forthcoming June general election, stating, "We challenge these three MPs to state clearly to the press and in their election propaganda what is their stand on nuclear power at Druridge. There comes a time when party politics and local issues conflict, and their constituents should assess which must take priority."

This was published in several newspapers, with attention-grabbing headlines. The *Weekly Courier*, delivered free to Morpeth, Wansbeck, Blyth Valley and North Tyneside gave our story front page headlines, CAMPAIGNERS' FURY AT MPs' VOTE. This newspaper was distributed throughout Neville Trotter's constituency.

His reply was, "My views are perfectly clear. I do believe that we need nuclear power in Britain, but I do not think we should have a nuclear power station at Druridge. What we want in the North East is a coal-fired power station. I believe we have to have nuclear power, but we do not have to have it on top of a coal field."

We had letters published in most local papers, pointing out how MPs' voted in the debate, and their views specifically on Druridge. Voters preparing for the general election were encouraged to challenge their prospective MPs on how hard they would be prepared to work for Druridge. Ideally, what the Druridge Bay Campaign would have liked was that every North East MP, regardless of political party, would be steadfastly on our side.

We wrote to all candidates in the three constituencies where the MPs had voted for Sizewell. Using their replies, we prepared illustrated briefing sheets, clearly explaining the candidates' views on Druridge, and how the sitting MPs had voted on Sizewell. Our active Support Group members then made door-to-door deliveries in Hexham, North Tyneside and Newcastle Central.

The election came and went. Druridge was only one of many issues upon which people voted. The election brought many more Labour MPs into power in the North of England and Scotland, even though it was yet again a Conservative victory.

For the three local MPs who had voted for Sizewell, the results were as follows.

Alan Amos stood in place of Geoffrey Rippon, who had retired. He increased the Conservative vote by 1000.

Piers Merchant lost his seat to Jim Cousins, Labour. His majority of 2228 turned to a majority for Jim Cousins of 2483.

Neville Trotter's majority of 9,609 in 1983 was reduced to 2,583.

Two of the three MPs had had their majorities significantly reduced, those who lived closest to Druridge Bay. We will never know how much our efforts contributed to this change, but we were pleased that it was obviously going to be more difficult than ever for any MP to favour nuclear development at Druridge.

No praise is too great for Alan Beith and Jack Thompson, who used every opportunity they could to state our case in the House of Commons.

As with the lobby exercise, we were finding out how to popularise our cause, and how to encourage politicians to support us. Our message was being taken far and wide by our MPs, and scarcely a voice could be heard in the North East favouring nuclear power. Although we were not a party-polical campaign, we had a firm backing in our region from local politicians. We had progressed from vaguely knowing that we had general support among local politicians to a refined knowledge of where our strength lay. It was substantial.

NOTES

1. *Journal*, 13 9 85
2. *Morpeth Herald*, 20 2 86

7. THE ARGUMENTS
1983 – 1989
Environment, Safety, Better Alternatives, Jobs, Nuclear Waste

From the beginning of the campaign of opposition to nuclear power at Druridge Bay, participants had to be prepared to argue their case, and to counter that of the CEGB.

This was not easy. Many people had a straightforward dislike of the nuclear power proposal, but whereas they would freely discuss their feelings with their family and friends it was quite another matter to stand on a public platform and speak with command of the subject.

We had to investigate the issues involved and take our information out to stalls in the street and various fairs and shows. It was often necessary to accept invitations to speak at meetings with a highly-trained and knowledgeable opponent from the nuclear industry putting the other point of view. We varied in our talents, but not in our dedication.

Toni Stephenson is a teacher, with no scientific background. She said, "If talking to people no more knowledgeable than oneself, you can appeal to their heart, their emotions, rather than give lots of facts and figures. When I did my first talk at Swarland WI, it wasn't facts and figures they asked for, it was about me. Why had I become interested in the Campaign? Among those women, it was much more on a personal level. However I would try to learn anything I didn't know beforehand, in an attempt to have information. The problem was that I would forget it all later."

Toni had been to one of the pre-Inquiry meetings at Hinkley, where she had deliberately spoken to Sam Goddard, one of the directors of the CEGB. "I knew what I was going to say about Druridge Bay, and why. Important people don't mean anything to me really. They're just doing a job. I'm not awed by them."

Jane Gifford is an artist and long serving member of the Druridge Bay Campaign, including terms as vice chair and chair. She explained how she handled speaking in public.

"First I made it clear that not being an expert did not exclude me from making decisions that were important to my life. And as with most women, I felt that common sense was more important than what so called 'experts' said was safe. It would be quite possible to talk about the dangers of nuclear power without having to go into technical detail, especially as my particular interest is how nuclear power fits in with a certain political ideology. It is a highly technical and centralised way of generating electricity, with its own private police force, out of the control of the people it is supposed to be serving.

"I recollect the leaked minutes from Downing Street, in which Mrs Thatcher was prepared to use nuclear power as a tool to break the miners. My arguments against nuclear power did not have to depend on being able to tell a Megawatt from a Gigawatt."

Gary Craig, DBC chair for two years also had no worries about addressing "experts". He said, "I had a degree in Maths and Physics. I was able to combine my political perspective with scientific training. And I wasn't overawed by the nuclear industry people. For example, I took part in a radio programme with George Johnston, CEGB engineer responsible for Druridge Bay. I could challenge him when he tried to be overtechnical."

One of the earliest moves we had made was to assemble a list of supporting "experts", who would be relied upon to help us with information, and in extreme cases be asked to speak on our behalf. They could not be expected to attend the less technically demanding type of meetings, such as at local women's groups or schools. Their time was too valuable. That kind of work had to be done by rank and file campaigners.

Friends of the Earth, in the group's early activities in 1983, had mustered their arguments under five main headings. They objected to nuclear Druridge for reasons as follows:

1. Destruction of a Rural Environment
2. Safety
3. No Need
4. Jobs
5. Nuclear Waste Disposal.

These categories were similar to those identified by the original Druridge Bay Association, although FoE had summarised them independently. They continued to form the basis of our arguments over the years. As our affiliated groups had often various reasons for opposing nuclear power at Druridge, it had to be pointed out at public meetings that these headings covered the full range of arguments. Not everyone was concerned about all of them.

How we dealt with these arguments, and the information we used, follows.

• • •

The first category listed above, destruction of the rural environment of Druridge Bay, was the most basic objection among our members. All campaigners agreed that this could not be tolerated, regardless of their views on other aspects.

As Hilda Helling put it, "At the heart of our campaign was the belief that the unspoilt beauty of Druridge was of value and benefit to everyone; that its pristine environment was a treasured part of our heritage; and that the threat to its future well-being posed by the CEGB's plans would be met with either disquiet or downright opposition in many dissimilar sections of the community."

The concern about the preservation and enhancement of Druridge Bay was so widespread, and so non-contentious locally, that we decided in 1987 that I could make a special slide show on the subject. This aspect of our work was very acceptable to wildlife groups, Women's Institutes and organisations who steered clear of anything remotely "political".

Speakers for the campaign found sympathetic audiences when presenting this argument, our first and most basic one.

The beach is a six mile curving stretch of sand, backed by mature dunes. Behind the dunes is farmland. Other long curving beaches lie further north, but Druridge is the nearest to Tyneside with no evidence of industry around it. It is spacious, and rarely crowded, largely because our north-eastern climate is often cool and windy. On milder days, which can be at any time of the year, families appear, dogs scamper, and racehorses have a six mile uninterrupted gallop on the sands.

However, Druridge Bay is in a mining area. Behind the dunes, and under the waters of the bay lies coal. There have been spasmodic attempts at mining this coal over the centuries. There were collieries at Widdrington, Radcliffe, Broomhill and Hauxley, now long closed. Since the war, large areas in the northern half of the bay have been systematically subjected to opencast mining. Adjoining the dunes, were three sites at Hauxley, Ladyburn, and Radar. East Chevington is currently still operating, though drawing its operations to a close.

These opencast sites were carefully landscaped to cause minimum disruption to the beach area during operation, and have now been restored. As part of the restoration programme, the National Coal Board, now British Coal, returned much of the land to farming and some of it to local organisations for nature reserves.

Therefore in the northern section of the bay there are extensive areas of dunes and hinterland which are now parks or nature reserves. Northumberland Wildlife Trust has nature reserves at Hauxley and Druridge Pools. Northumberland County Council has a three mile stretch of dunes and large hinterland area with a lake, at Druridge Country Park. The National Trust owns a mile of the mature dunes area. After East Chevington opencast site is finally restored, Northumberland Wildlife Trust has a negotiated agreement to develop another nature reserve, with reedbeds.

Towards the southern end of the bay there is a substantial east-west fault, called the Grange Moor Fault, which divides the bay in two sections as far as mining is concerned. Deep mining from Ellington Colliery, in the south of the bay, comes to an end at this fault line. There is a small area of land north of the Grange Moor Fault, squeezed in between that and the large opencast sites to the north of the bay, which has neither been opencast nor deep mined. The CEGB's test drillings showed them that this area is stable. This is Hemscott Hill farm, 300 acres of which they have purchased. It is the only part of the six mile stretch of the bay which is suitable for nuclear power stations.

The Grange Moor Fault runs east-west through Hemscott Hill farm, and the CEGB did not purchase the southern section which had been undermined.

South of the fault, the effects of deep mining produce a different landscape from that of the northern section of the bay. Colliery subsidence has caused the land to sink. Marshes and ponds have formed where drainage is poor. These freshwater ponds, with some tidal input of brackish water from the sea, have become important nature reserves, and at Cresswell Ponds, are a Site of Special Scientific Interest (SSSI). Northumberland Wildlife Trust, in co-operation with British Coal and Alcan Farms Ltd, manage Cresswell Ponds.

Mining has thus had its effects on Druridge Bay. Yet careful management has enabled the landscape to be recovered, and in some ways even to be enhanced.[1]

Castle Morpeth Borough Council owns a half mile of dunes in the south of the bay. Most of the bay lies within Castle Morpeth Borough Council's administration area, with a small northern section lying with Alnwick District Council. Both councils are encouraging sensitive development of these areas, for leisure and recreational rather than industrial use.

In summary, the past effects of industry are being erased at Druridge Bay, and the last thing needed is the disruption caused by a massive civil engineering project of one or more nuclear power stations. Local councils and conservation groups have been working hard to improve the hinterland of the bay. Today the visitor finds birds, wild flowers, grazing animals behind the dunes, and over them the sparkling panorama of the beach itself, with its ever changing patterns and colours of sea, cloud, waves and sand.

• • •

The second topic campaign speakers had to cover was that of safety. When the Druridge Bay opposition first became active, the Chernobyl accident had not happened. Campaigners would point to the near meltdown at Three Mile Island in the USA in 1979. This was worrying to many people, especially as it was the same type of reactor as that proposed for Sizewell and the rest of the British sites including Druridge Bay.

Safety is a difficult subject for non-specialist lay people to tackle, including myself. One has to understand some physics and engineering.

The early Druridge Bay Association used Walter Patterson's classic primer for nuclear campaigners, "Nuclear Power". He said, "Easily the most controversial feature of the PWR is the emergency core cooling system, provided to prevent over-heating of the reactor core in the event of an accident."

FoE Mid-Northumberland produced a tape-slide show in 1985 which dealt with the difficult safety problem by interviewing experts. Member and Newcastle Polytechnic chemistry lecturer Dr Ed Metcalfe gave a lucid explanation of how the cooling system might fail.[2]

The tape-slide show ran by itself, and there were not often problems as our audiences were usually no more knowledgeable than we were. It was our principle to assure them we were grappling with the same problems that they were, and that we could take queries away to our expert advisors.

A more easily presentable piece of information concerned the emergency evacuation at Three Mile Island, following the near-meltdown. Tyneside Anti-Nuclear Campaign had produced a map of the north-east, with various circles surrounding Druridge Bay superimposed upon it. The circles included actual evacuation zones at Three Mile Island. At up to a 5 mile radius from the damaged

American nuclear power station, pregnant women and pre-school children had been evacuated. Up to 10 miles, schools had been closed, and the population told to stay indoors. Up to 20 miles, the population had been told to stand by for evacuation. The outer 20 mile circle in our own case included all the Tyneside conurbation. It helped people to see the enormous problems of evacuation.

The CEGB however did not think it was necessary to prepare for evacuation further than 1km from the nuclear power station. In the previously mentioned Nuclear Power Investigations prepared for the people of Northumberland by the County Council in 1979, the CEGB declared, "In all credible fault sequences, regardless of weather conditions, the release of radioactivity would **not result in any need for action to be taken off the power station site.** However, it is considered prudent to have contingency plans which allow for the temporary evacuation of any people living very close to the power station. **The contingency plans... cover the area around the site to a distance of 1km** nominal radius." (Author's emphasis.)

We had no trouble convincing people of the dangers of nuclear power after the Chernobyl accident in 1986. We were able to use the report on the consequences of a nuclear accident at Druridge Bay, which had been done for us by London scientist Dr Charles Wakstein, to make a series of information leaflets. These gave evacuation figures and potential death rates for the centres of population across the north-east and quoted the extremely varied accident estimates from "expert" bodies.[3]

Dr Wakstein said, "Even if the odds of Chernobyl happening were once in a thousand million years, it **has** happened, and there is a growing conviction among engineers and scientists outside the nuclear establishment that any technology in which an accident can have consequences as far-reaching as those calculated for a Druridge Bay Meltdown ought to be politically unacceptable."

The importance of our monitoring work was increased by the Chernobyl accident. The safety arguments spoke for themselves in the years 1986 and 1987. They have been superseded in the public's mind for the time being by later developments on the economics of nuclear power, and because there has not been a serious nuclear accident since Chernobyl.

● ● ●

The third topic which we had to argue was whether or not we needed nuclear power for secure electricity supplies. It is easy enough to make vague statements that there is plenty of coal, or that renewable energy is preferable, but people need to be convinced that these methods can provide sufficient electricity. In recent years, it has become increasingly important to include environmental considerations in our argument, such as the effects of acid rain and the Greenhouse Effect.

When nuclear power at Druridge Bay was first proposed, over eighty per cent of British electricity was generated by coal-fired power stations. Local coal contributed to this, being burnt in the region and also in the south-east of England. The suggestion that nuclear power might take over some of this role caused an examination of the future use of coal.

By January 1981, some members of Northumberland County Council's planning committee had begun the call for investigation of a further coal-fired power station at Blyth as an alternative to a large nuclear power station. The CEGB owned enough land beside the two existing power stations for the construction of a third.

Ellington Colliery was producing about 2 million tonnes of coal per year and there were reserves of up to two hundred million tonnes in the area north of the colliery, under the sea. What was not known was whether or not the National Coal Board intended to put the investment required into mining these reserves, and whether it would be available for local coal-fired power stations.

In 1979, the CEGB had estimated that a new nuclear power station would be needed in 1990. They estimated that electricity demand would increase in the North-East Coast Area between 1979 and 1990/91.

Campaigners had to answer the question as to how our electricity could be supplied, if not by nuclear power at Druridge Bay.

The Druridge Bay Association in 1979 had put forward the idea that any new power station that was absolutely needed should be built to provide combined heat and power. By using waste heat for domestic or industrial purposes rather than letting it escape into the environment through cooling towers or into the sea, these power stations use 80% of the energy value of coal, compared to about 35% from a conventional station. These power stations are clean, very economical with fuel and small enough to be built in urban areas.

More detailed information was needed on the full range of alternatives to nuclear power, and during January and February 1984 Friends of the Earth set up a series of lectures in Morpeth. Dr Nigel Mortimer of Sunderland Polytechnic gave us an overall view of energy alternatives, both in energy generation and conservation. He explained further the advantages of combined heat and power, which is more developed in Scandinavia than in Britain. Half a decade before "energy efficiency" had become a household term, Dr Mortimer was recommending the advantages of many methods of reducing electricity consumption including the use of low-energy light bulbs.

In October 1985, he followed this with research which summed up ways in which the north-east could be self sufficient in electricity supply, after the year 2000, without the need for nuclear power. This we were able to convert into useful information for ourselves.[4]

Dr Nigel Mortimer continued to be one of our most valued experts. He explained why he helped us in this way. "I think nuclear power is a very dangerous and wasteful distraction from the real energy issues. Academics should serve the community when able, especially when such developments are on their own doorstep. Also, I don't like scientific arrogance, as epitomised by the CEGB."

We adopted Dr Mortimer's arguments for alternatives to nuclear power at Druridge Bay. We produced a simple leaflet based on his work. In May 1989, detailed, up-to-date research became available in the form of an EC sponsored energy study, the North-East Energy Study, to which Dr Mortimer was one of several contributors. This caused certain of his early conclusions to be modified, but his work served as a scholarly and authoritative basis for our case for several years.

• • •

The fourth topic to be argued was that of jobs. There was a high unemployment rate in the north-east. In the Morpeth/Ashington area it rose from 16% in 1985 to 22% in 1986. The CEGB were aware of that, and tried to take advantage of it in their propaganda.

There was a certain amount of faith in 1979 and 1980 that jobs would be available for the local unemployed with any future construction of nuclear power stations at Druridge Bay. Dave Black, at that time reporter for *Northumberland Gazette*, remembered, "One day back in 1980, I was accosted outside Bedlington magistrates court by two local 'heavies' who had seen me emerge from a vehicle with the *Northumberland Gazette* title emblazoned along its side. They wanted to know when the Press was going to print some stories extolling the virtues of the proposed nuclear power station at Druridge Bay, which, they were adamant, would provide much needed jobs for people like them. I cannot provide you with a verbatim note of their comments about the Druridge Bay Campaign, but suffice it to say that you could not use them as cover quotes for the Druridge story, which I presume is meant for family reading.

"The episode was a salutory reminder that the argument is not entirely one-sided, and that for some people the potential threat posed by nuclear power is a price worth paying for what they perceive as the more important short-term benefits of a job and more cash in their pockets".

In the widely distributed publication of December 1980, the CEGB claimed that jobs for construction workers would peak between 1500 and 2000. By the time they published their Druridge Brief, in May 1984, this had risen to "employment for over 3000 workers would be provided during construction". They also said construction would take "some seven years". Nowhere did they make it clear that there would be far fewer than 3000 jobs for most of those years, that they were only temporary, or that most of the workers would be brought in from other parts of the country. We found these things out for ourselves.

SCRAM had published a booklet, *Torness – from Folly to Fiasco*, which revealed that 3000 jobs had been promised to construction workers in the Torness project. It was helpful to be able to use the experience of earlier campaigners who had confronted the same situation and exposed its fallacies.

The booklet related that in February 1981, John Home Robertson MP East Lothian had said in a Borders newspaper, "The promise of jobs at Torness is the biggest let down of all. At the last count we still had 667 construction workers on the dole in East Lothian and Berwickshire, many of whom have been turned down for jobs at Torness. Meanwhile for reasons best known to themselves, the contractors at Torness are employing 718 workers from outside the Lothian and Borders region."

By February 1982, 3120 workers were employed in the construction of Torness. Of these, 2024 or 64.8% were from beyond the East Lothian, Borders and Edinburgh region.

We visited John Home Robertson in 1984, and he confirmed the above. He agreed to be interviewed for Friends of the Earth Mid Northumberland's tape/slide show.

As well as discovering that the net gain to local construction workers was minimal, we also found out that other local jobs already existing were threatened.

John Home Robertson pointed out that in his area there were 5000 people employed in mining and in the local coal fired power station, whose jobs were threatened. The parallels to our own situation were impressive, with nearly 7000 men employed in local collieries and Blyth coal fired power stations.[5] Coal from Northumbrian collieries went to Blyth, as well as to power stations in the south east. Changing the emphasis from coal-fired electricity to nuclear could lead to large job losses.

Dr Mortimer, with Gary Craig and Martin Spence of Tyneside Anti-Nuclear Campaign's Trades Union Group estimated that 5000 miners' jobs were threatened, and in associated industries a further 2000 including power station workers, railway workers, port employees and seamen. They pointed out that construction of a third large coal fired power station at Blyth would maintain these local jobs, and create more in construction.

It became even easier to disparage the CEGB's promises of jobs after a Jobs Forum which the Druridge Bay Campaign organised in October 1987.

Speakers were invited who could offer specialist experience in the effect of nuclear power on local employment. Campaigners and councillors assembled in Gateshead Civic Centre's Council Chamber, lent to us for no charge for the day by Gateshead Metropolitan Borough Council.

Expert speakers were Tom Miller, a planner from Castle Morpeth Borough Council, Dr Brian John who had studied employment round Trawsfynydd nuclear power station in North Wales, Cllr Will Herald of Lothian Regional Council who gave more information on employment at Torness, Dr Eric Wade, Open University, who outlined risks to the coal industry, Barney McGill, who described problems faced by the power engineering industry on Tyneside and Dr Nigel Mortimer who talked about employment on alternatives to nuclear power.

One of the most influential of the papers was that produced by Tom Miller of Castle Morpeth Borough Council, the local authority in which the Druridge site was located. At the Druridge Bay Campaign's request, Tom Miller, council planning officer, had agreed to summarise a part of a CEGB sponsored study by Oxford Polytechnic called *The Socio-Economic Effects of Power Stations on their Localities.*

The report dealt with employment impact of construction and operation of nuclear power stations. He concluded that of the high number of workers needed for construction over half would be imported labour. Of "local" labour, many workers would come long distances, from as far as Carlisle, Middlesborough and beyond Berwick. Many of these jobs are not long lasting, and for the semi-skilled or unskilled.

Once the power station was in operation, a large proportion of the highly skilled personnel would come from outside the local area. Perhaps 200 local jobs would be available, in semi-skilled or unskilled grades. These local jobs would include a 10 – 15 mile travel to work distance, and hence any impact would be hardly noticeable in the Castle Morpeth area.

Tom Miller summarised, "The conclusion that one inevitably reaches is that despite the vast sums expended on construction, the undoubted amenity losses of such large structures on the local countryside and the perceived disruption to life by local people, the actual jobs gains of large structures such as power stations are not all that great."

He compared the employment impact of nuclear power at Druridge with that of Searle Pharmaceuticals, near Morpeth. "It is basically a medium sized pharmaceuticals factory. It doesn't have a great impact on the local economy, and yet it employs about as many in total, both of indirect and direct jobs, that a nuclear power station at Druridge would be predicted to employ, ie 600 – 650."

Although a Castle Morpeth Borough Council officer, Tom Miller had prepared this work for the Jobs Forum in a voluntary capacity in his spare time. He is now Head of Planning in Ellesmere Port and Neston Borough Council, South Wirral. In a recent letter to the Campaign, he told us the following:

"I tried to approach writing the paper with an open mind, neither pro- nor anti-nuclear power station. In the end, I was very surprised at the low levels of employment that would be generated by the nuclear power station once it was built. It would have added relatively little to the local economy."

Cllr George Ferrigon of Wansbeck District Council was particularly impressed by Dr Brian John's talk. Wansbeck District Council's populated area is only 6 – 7 miles from Druridge Bay. He recalls, "Dr Brian John explained that in the initial years there was new work, but in the long term unemployment went up. People in North Wales were left with very few long term jobs at the nuclear power station, but once it was built they were saddled with it. Besides that, there is the problem of attracting new investment, and the nuclear power station took jobs away. Investors are reluctant to move next to it because of reaction from their staff, and if several different sites are available another will be chosen. In the Wansbeck District, we have enough problems attracting new industry without having a nuclear power station on our doorstep."[6]

A letter from John Home Robertson to the Campaign in July 1988 confirmed the difficulties in the Torness region, after the building of the nulcear power station was complete. He wrote, "We are now faced with the problem of reconstructing the local economy of the Dunbar area, and trying to find employment for a lot of people who came into the area to work at Torness, as well as those local people who were lucky enough to get employment there."

All this information enabled us to put together a very convincing case against the CEGB's casual claims for thousands of local jobs.

Cllr Janet Cann said, "It was particularly useful for the planning authority Castle Morpeth Borough Council to have this information. The council wanted to generate employment in that part of their borough, round the high unemployment blackspots of Amble and Hadston. The research, particularly that of Tom Miller, became a well-argued case against nuclear power stations as a major employment generator. It was most useful when talking to other councils, particularly Alnwick, which like ourselves had an unemployment problem in that part of the county".

The DBC had helped to show that the employment benefits claimed by the CEGB were greatly exaggerated. When weighed against what we stood to lose, they were negligible.

• • •

The fifth important reason for opposing nuclear power at Druridge Bay was the problem of the disposal of nuclear waste which is active and dangerous for hundreds of thousands of years. While arguing the case against nuclear power at Druridge, we re-inforced the public's realisation that there was no solution to this problem. Where would the waste from Druridge go? In some deep hole far away, or out into space, or into the Cheviots?

The saga of nuclear waste disposal consists of one episode after another of resistance by local people. [7]

In Northumberland County Council's 1979 *Nuclear Power Investigations*, one of the questions asked of the CEGB was "How would solid wastes be disposed of?" The answer provided was that "the irradiated fuel would be transported to British Nuclear Fuels Ltd (BNFL) plant elsewhere for reprocessing and storage of the resulting high level waste."

The BNFL plant was Sellafield, but the ultimate destiny of the waste was not mentioned, for the simple reason that it was not known.

In the *Druridge Brief* of May 1984, the CEGB put a supposedly typical question from the public to themselves as follows. "How will you dispose of the waste products?" Their answer was, "Used fuel rods would be transported by road and rail in massive steel containers, known as flasks, probably to Sellafield for reprocessing."

There is no mention of what reprocessing is; the inference which could be drawn by the layperson is that it means dealing with it in some satisfactory way. There is no mention of where the products of reprocessing go, or whether they would stay at Sellafield.

Clearly, people who were opposed to nuclear power at Druridge Bay were supposed to be satisfied that the waste would be taken away from the area, and would no longer be a local problem. Perhaps that was all the CEGB cared about. They paid BNFL enormous amounts for taking the spent fuel off their hands.

As with having to argue on safety issues, discussion of nuclear waste requires campaigners to have a certain knowledge of chemistry and physics. This was daunting for many of us.

Friends of the Earth in London have an efficient Energy Campaign and have been a reliable source of information and advice to us. We used one of their publications, *Radioactive Waste – The Gravediggers Dilemma*, to make a slide, showing that sending spent fuel to Sellafield increased the bulk of nuclear waste by nearly 200 times. This vastly increased the ultimate disposal problem. Thus nuclear waste from Druridge would certainly multiply the waste disposal problem if sent to Sellafield.[8]

But it was not only nuclear waste produced during electricity generation that was the problem. Disposing of the enormous radioactive mass of the nuclear power station at the end of its working life was another dilemma. It was vaguely glossed over in CEGB propaganda. They did say in *Nuclear Power Investigations* that it would be "advantageous to delay its dismantling for some 50 or more years after shutdown" and that "studies are being carried out to determine at what stage final dismantling might take place."

In May 1984, in a letter to the local papers, we exposed the CEGB claim that "long term delay in dismantling" would be attractive. A CEGB spokesman had said, "The buildings would continue to be supervised for 100 years or more."

We pointed this out to an always astonished public wherever we were invited to speak. People were amazed to find out that the CEGB actually intended to seal up and guard a radioactive building possibly 50 metres in height for at least 100 years. That extraordinary plan applies, now, to all the British nuclear power stations dotted around our coastline.

When one considers the social and political changes that have occurred in the last hundred years and what might happen in the next hundred years, it seems absurd to the point of insanity. What if there should be a "conventional" war in which the seal should be broken? Suppose the nation could not afford the cost of containing the radiation? What if the political will to protect the public should cease to exist?

On certain occasions, when feeling despondent at the difficulty of proving a point over the CEGB men, who were fond of criticising "emotional" arguments, I would remind myself of their proposals. These people really believed it made sense to cement up radioactive sites for a hundred years, and leave them for future generations to deal with. It was clear to me whose point of view was more rational.

The report of the House of Commons Select Committee on Energy published in June 1990 examined the cost of nuclear power. Ten years after the first CEGB publicity brief on nuclear power at Druridge Bay, it had this to say about nuclear waste disposal and decommissioning of nuclear power stations.

"Dealing with spent fuel involves potentially open-ended liabilities, both for reprocessing and for final disposal of waste...The cost of final disposal of waste is the least well-defined of all nuclear costs, since no final repository for nuclear waste has yet been selected."

"No commercial nuclear reactor anywhere in the world has yet been fully decommissioned and there is no experience with the potentially difficult stage of dismantling and disposing of the most contaminated parts of the reactor. The risk is both of increases in cost and of the regulatory authorities not allowing utilities to wait 100 years before ultimate de-commissioning."

Today, now that the full costs and problems of decommissioning are clearer, it is easy to be caustically critical of the evasiveness of the CEGB. But in the early 1980s, we campaigners were an assorted group of concerned citizens with little specialised knowledge. We were struggling with a difficult topic. The dark-suited "establishment" figures of the CEGB carried an air of authority on public platforms, and it required courage to stand up to their often sneering and patronising attitudes. The Select Committee report was evidence that our cyncism had been justified.

NOTES

1. Another less obvious effect of mining is its contribution to coastal erosion. A report commissioned by Castle Morpeth Borough Council in 1985 concluded that severe coastal erosion, particularly in the northern half of the bay, was probably due to several factors caused by mining. Firstly, although mining engineers tried to ensure that sea defences were not subjected to undermining, there were places where this did happen, causing subsidence. Secondly, under-sea mining has caused subsidence of both the sea bed and the rocky outcrops, or scars, at the north and south ends of the bay. The scars help to lessen the force of the waves. Deeper water results in stronger waves, which act on less effective scars and causes erosion.

2. "One of the main problems that can arise with the American PWR is from a leak in the cooling water circuit leading to a loss of pressure, known as a LOCA or loss-of-coolant accident. This is what happened at Three Mile Island, where a pressure relief valve jammed open. As pressure falls, the cooling water is converted to steam, a far less efficient coolant than water, so the core temperature rises. The Sizewell B design contains several safety devices for emergency cooling for the core. These systems may fail for a number of reasons, such as mechanical damage, or they may be misused by the operators as happened at Three Mile Island. The temperature of the radioactive fuel could shoot up several thousand degrees, even if the reactor SCRAMS. (SCRAM is a code word for the process which terminates the reaction.) The control rods are driven into the core to terminate the nuclear reaction. Without cooling, enough residual heat is generated to initiate melting of the fuel in about one minute.

"As the temperature rises, the steam would react with the zirconium fuel cladding, releasing explosive hydrogen gas. A hydrogen explosion was recorded at Three Mile Island inside the pressure vessel. Eventually the entire core of molten fuel and radioactive waste would melt its way down through the base of the reactor and into the earth below, with a release of large amounts of radioactivitiy into the environment. By this time it would be necessary to evacuate the area." Dr Ed Metcalfe. FoE interview, 1985.

3. The CEGB at the Sizewell Inquiry said that a reactor meltdown could be expected about twice in a thousand million years.

The US Rasmussen report found a 50/50 chance that a major accident could occur every 23 000 reactor years.

Die Deutsche Risikostudie found a 50/50 chance that a major accident could occur ever 10 000 reactor years.

These above are "technical risk assessments".

Islam and Lindgren, *(Nature* 21.8.86), used data from actual operating experience. There had been two major accidents, Three Mile Island and Chernobyl, in 4000 reactor years. They therefore calculated a 70% probability that one accident could happen in the next 5.4 years. The probability of having one accident every two decades is more than 95%.

4. Coal would remain the largest single contributor to north-east energy needs, at about 52%. This would supply direct heating and combined heat and power, as well as some synthetic fuel. Geothermal reserves in Weardale could provide heat plus electricity to 33% of the total. Biomass burning would provide 12.5% of our electricity. Small contributions from wind, solar and hydro power brought the total up to 100%

The figures were based on 1984 energy demand, and did not include the very great potential reduction by conservation. Because of the inevitable delays in implementing these non-nuclear systems, a new, large, coal fired power station would be needed to fill the gap. This should be fitted with appropriate pollution controLs, and could possibly be built at Blyth. Dr N Mortimer, *Regional Energy Policy: Renewable Energy Sources for the North-east*, October 1985.

5. In the House of Commons in February 1982, he had said, "There is in my constituency a 1200 MW coal burning power station at Cockenzie, which burns virtually the whole output of two coal mines at Monktonhall and Bilston Glen, as well as coal from a nearby opencast site. The power station employs 600 people. The two collieries employ more than 4000 miners. It follows therefore, that Cockenzie power station accounts directly for the employment of nearly 5000 people".

In 1985, Blyth A and B Power Stations, with a total capacity of 1700 MW, employed 800 workers and there were 6000 miners employed in Northumberland.

6. Dr Brian John, described the impact of imported workers who remain in the local area when construction is completed. He pointed out that local unemployment rises, as many men stay behind and go on the dole. During the construction phase, local authorities must provide extra services, money for which could have been spent more usefully elsewhere.

He explained that fifty per cent of the local workforce moved sideways from other jobs, hastening the decline of other industries. The relatively high wages of those employed in the power station caused an inflationary effect in the area, with pressure on other local employers to raise wages. He pointed out that a sterilised zone tends to develop round the power station, where building projects are not carried out. This could vary from about 5 to 25 miles, depending on public perception of the risk. He concluded that in Wales, a mere 158 long term jobs had been gained, other than would have been developed anyway. For further details, see *Nuclear Power and Jobs: The Trawsfynydd Experience*, 1986, by Dr Brian John.

7. Outrage over the choice of the Cheviots and similar sites for high-level waste disposal caused the government to postpone making a decision on this unpopular topic in December 1981.

NIREX (Nuclear Industry Radioactive Waste Executive) was set up in 1982 by the government to try to find a solution for disposal of low and intermediate-level waste.

In 1983, the former Billingham salt mines, under ICI in Cleveland, were suggested as a possible choice for an intermediate-level waste dump. An outraged public and well organised opposition, Billingham Against Nuclear Dumping (BAND), convinced ICI not to co-operate with NIREX, and the proposal was dropped in 1985.

Currently two sites for the disposal of intermediate level nuclear waste are being considered, near Dounreay in Caithness and under the sea from Sellafield in Cumbria, with the latter the favourite.

In October 1983 four sites on the clay belts closer to the electorally sensitive south east were suggested for the disposal of low-level nuclear waste in Lincolnshire, Bedfordshire, Essex and Humberside.

As was to be expected, local people in those areas set up effective campaigns to protect themselves. In 1986, the four sites joined together to form BOND, Britain Opposed to Nuclear Dumping. In May 1987, the sites were dropped by NIREX, just one month before the pending general election.

8. The question arises as to why the waste should go to Sellafield at all. The official purpose of reprocessing is to extract plutonium and unburnt uranium from the spent fuel, for future use. With regard to plutonium, its only significant use is for the fast breeder programme, which has effectively been abandoned in the UK. Plutonium from the civil electricity generation programme is not allowed to be diverted for military purposes. With regard to uranium, it is far cheaper to buy freshly mined and processed uranium than to reprocess spent fuel at Sellafield. Why then is this expensive and polluting process carried out, which vastly increases the problems of nuclear waste disposal? What would be the ultimate destination of spent fuel from Druridge Bay? These are questions to which so far there have been no satisfactory answers.

National Friends of the Earth's position on reprocessing is that it should stop. No new nuclear power stations should be built, and existing ones closed. Nuclear spent fuel already in existence should be stored on the site where it is produced, in dry storage plants. This would at the very least prevent the vast increase in bulk caused by reprocessing, and prevent the need for having to find more sites to be contaminated by storage of extra radioactive materials.

8. WHAT PEOPLE HAVE DONE
1983 – 1989

For over twelve years, a steady stream of people has contributed to the work of the Druridge Bay Campaign. Some have come and worked for a while, then moved on. Others have stepped in to take their places. Some people have worked for the Druridge effort for many years, and are still doing so. Certain people took on humble jobs for which there was little outward reward, others took the limelight. It would be quite impossible to mention all these people by name.

Friends of the Earth Mid-Northumberland had established the principle as far back as 1983 of everyone using whatever skills he or she has to help save Druridge Bay; every person has something to offer, and no skill is too humble.

During the active phase of Mid-Northumberland FoE, between 1983 and 1985, supporters from outside the group, with specialised skills, were willing to help as well. Many of these people were talented professionals.

Roy Beasley, principal of Ashington Technical College was a member of a quartet called "The Stewart Singers". He had attended the Cairn ceremony, and later offered to organise a fund-raising concert. Together with Shirley Wilkinson soprano, Jean Kelly contralto, Keith Stewart tenor, and Gillian Stewart on the piano, he planned a programme of music, poetry and prose readings on the theme of care for the countryside. Friends of the Earth booked Morpeth Town Hall Ballroom for 11 February 1984. Alan and Barbara Beith, and Jack and Margaret Thompson, attended. The Liberal and Labour MPs affirmed their unity on opposing nuclear power at Druridge Bay. It was a pleasant musical evening and an opportunity for fundraising and publicity.

Encouraged by our first attempt at a musical evening, we approached Martin Shillito, French horn player in the Northern Sinfonia orchestra. I had met him and his family while camping in Northumberland the previous summer. He agreed that he and some of his musician friends would organise a concert on Friday 25 May 1984. The musicians were Sylvia Sutton violin, Julie Monument violin, Andrew Williams viola, Jeanette Mountain cello, David Munroe double bass, George McDonald clarinet, Roy Thorndycraft bassoon and Martin Shillito French horn.

Before the concert Martin Shillito told the media, "We are very concerned about the possible wrecking of this unspoilt area. I can remember seeing Windscale as a child, which has now done its damage, and is continuing to do more. I don't want to see the same thing happen in this area. The environment should be protected for the sake of our children."

The concert was preceded by the arrival of the drilling rigs on May 17th, described earlier. People's attention was therefore very much on the Druridge issue. The concert was an inspiring event during a distressing time for campaigners. Drilling into the land at Druridge Bay was taking place while the music was being performed.

Another helper during that summer of 1984, as Friends of the Earth were trying valiantly to raise £700 000 to buy the land, was Hilary Town, a Bedlington potter. She offered to make a series of commemorative mugs for sale. This offer was gladly accepted. She said, "They show in relief a picture of the cairn at Druridge Bay, with the bay itself going around the side of the mug."[1]

Next to help were the Addison Rapper and Clog dancers. They offered to perform in March 1985. FoE organised a ceilidh in Morpeth Town Hall, and Addison dancers did a cabaret spot. They invented new "Dances for Druridge", which were to become a standard part of their future repertoire, and they performed them for the first time that evening. Before and after their cabaret spot, they toured the Morpeth pubs, giving performances, and shaking a collection box.

Another important person to offer his skills was Ray Scott of Ellington. He was a keen runner, and had trained on the beach and dunes at Druridge for many years. He offered to organise a Beach Run. Ray said at the time, "The beach at Druridge and the Cheviots are the two places I do my training. The nuclear industry threatens both environments." He organised some of his friends from Morpeth Harriers to help with supervising the run, while Friends of the Earth attended to publicity, collected entry fees and sponsor money, and organised refreshments. Ray and his wife, Hazel, and sons, Paul aged nine and Jonathon

Beach Runners, 1987. *Photo Newcastle Evening Chronicle and Journal Picture Library*

aged five, took part in the run. A photo of him with his family earned us publicity on the Sports Page of the *Weekly Courier.*

Other experts added to the publicity we got from the Beach Run. Amber Films from Newcastle wished to televise the run, as part of a film they were making for Channel 4. They posed in photos for the local papers.

Our publicity also attracted the attention of performers in the Royal Shakespeare Company, on their spring schedule in Newcastle. Actresses Katharine Rogers, Josette Simon and Kaye Buffery organised their co-performers into running. Katharine Rogers described her reasons. "A lot of us feel strongly about preserving the landscape and Druridge is a beautiful area. About twelve people from the cast have promised to run, and they have got quite a bit of sponsorship from company members. Brian Blessed, who is not running, is sponsoring us for £50.00."

Each time celebrities came forward, the standing of our campaign was enhanced. We took every opportunity to publicise their participation.

Our efforts to draw in the whole community, whatever their interests, were summarised in a DBC quote to the newspapers at the time of the Beach Run: "This event is organised to draw in yet another sector of the public. We have had a cairn laying ceremony for VIPs, a family style Northumbrian Gathering, concerts both classical and lighthearted, a protest demonstration and petition drives. We hope to give everyone in the area a way to express their opposition. This time it is the turn of the athletes."

It can be seen that the principle of involving as many people as possible was well established by the time the Druridge Bay Campaign took over the momentum from Friends of the Earth Mid-Northumberland in early 1986.

From 1986 onwards, the DBC became an increasingly large and complex organisation. It required management skills, and good leadership. A series of officers including treasurer and membership secretary was required. Funding had to be obtained and supervised. Once the decision to employ workers was taken, the DBC's managers became employers, with the attendant responsibility.

The Druridge Bay Campaign's first chair was Gary Craig. He had experience in trade union work locally, and strong principles of opposition to nuclear power. He helped establish the Druridge Bay Campaign as an organisation which had room for the full range of opposition, including groups like the National Trust and councils with Conservative majorities. He also had many contacts among trades unions and councils. This was very helpful to the DBC. He was an excellent chair of meetings, keeping the issues in hand firmly, yet allowing all viewpoints to be expressed and the wish of the meeting to prevail. He was also a considerate employer. These qualities set the tone for the future of the DBC.

After two years as chair, Gary resigned in 1986. I felt at the time that there would be no-one who could do the job so well. However, Gary always claimed that new talent would emerge if given the opportunity. He had seen his job as helping to lay the foundations for a long struggle for Druridge Bay.

He was succeeded by Jim Wright, representative on the DBC Steering Committee for Ellington and Linton Parish Council. Jim felt it was a great privilege and honour to take the position. He said, "It was particularly valuable to meet members of other political parties, and it was good to see how much common ground there was between us." As an active member of the Labour Party, Jim was able to inform and even influence members on DBC developments.

Jane Gifford became vice-chair at this time. This was a role she took until October 1990, when she was elected as chair. Jane had vast experience in work opposing nuclear power as a member of Tyneside Anti-Nuclear Campaign. She had been involved since the Torness demonstrations and the Cheviots Inquiry. Her job as vice-chair of the Druridge Bay Campaign involved much close collaboration with the daily activities of the workers in the office. She worked part-time as an art lecturer and was available in the daytime, which was an asset. She became very skilled at campaign management over the years.

In October 1987, Jonatha Robinson of NALGO was elected to the post of chair, which she occupied for two years. Jonatha was very keen to help guide the DBC close to its remit, which was to concentrate on being opposed to nuclear power at Druridge Bay. She sometimes felt that the DBC was perceived too much as involving those whose lifestyle was "alternative". She saw it as part of her job as chair to stick to the straight and narrow track, to have a campaign where people of all views, all lifestyles and all political persuasions could rally to the cause of saving Druridge Bay. The single issue was the DBC's strength.

Jonatha worked hard as chair for two years. It was a tremendously time-consuming job as the DBC expanded, and its programme of activities became ever more hectic. She portrayed the image of a sensible person from the grassroots. Jane's more radical and imaginative stance was a good counterbalance, and the two worked together with mutual respect and liking over the period between 1987 and 1989.

The job of treasurer was an arduous one. Clare Barkley was an administrator at Northumbria Water Board, an active member of NALGO, and a member of Mid-Northumberland Green Party. She had worked closely with Gary Craig at the fundamental organising of the DBC, including the setting up of the funding applications for workers. She took on the job from the early days in 1985 until September 1987. Clare was highly efficient, and her work for the DBC was much admired. In the end, the work of treasurer proved too overwhelming for someone in full-time employment. Sometimes Clare had to spend all day Sunday on DBC matters. When she resigned, she was a great loss.

For a short time, we had a series of temporary treasurers, until Gordon Steel took the job. Gordon was the representative on the DBC Steering Committee from Camperdown Labour Party. He had been made redundant when Wills tobacco factory closed at Wallsend. He was a magistrate, and devoted much of his free time to court attendance. He fitted in his work as treasurer with his many other interests such as playing the violin, cycling and working on racecourses at Gosforth and Hexham. Gordon was a familiar visitor to the DBC office wearing his motor bike gear, often on the way to climbing Cheviot or other outdoor activities.

He stayed as treasurer from September 1987 to September 1990, preparing wages, bills, balance sheets and the thousand and one details that needed attention.

Another campaign stalwart, Ron Major from Bedlington, took on the job as membership secretary in 1985. For over five years, Ron has dutifully posted out packages of materials to new members, and passed records in to the office, where they are processed by Wendy Scott. Ron is the Liberal Democrats' representative on the Steering Committee.

An essential and time-demanding job was dealing with what we called "the funding proposal". Fifteen councils were affiliated to the Druridge Bay Campaign, and paid a £50 annual fee. They also sometimes gave grants for specific requests. However six gave larger grants for the running of the office and worker salaries as described earlier. These councils had to be sent new applications for funding every year. Each council had its own particular timetable and routine for this process, so this was no easy job. We were very relieved when it was taken over by Betty Greenwell from Warkworth. She had retired from a successful career with IBM. Under her management, the applications were made and the money duly arrived on the office desk.

Whenever reports were required, particularly on strategy matters, another of our skilled volunteers would take on the job. Adrian Smith is a planner at Newcastle City Council. He was that council's representative on the Steering Committee, and had also been instructed to give assistance to the Druridge Bay Campaign as part of his job. He was experienced in planning matters which related to energy, and involved in Newcastle City Council's plan to set up a combined heat and power station. He could be relied on for cool logic, and advice on energy and strategy. He produced many reports for us in local authority style, with sections and sub-sections neatly delineated.

When leaflets and information for the public were required, the job usually fell to me as part of my employment. I was however greatly assisted by people to whom I could turn for advice and information. Jane Gifford, as an artist, would provide illustrations, often witty ones, to request. Many times during meetings, Jane would sit with a well-chewed stubby pencil end and produce the drawing that was needed for the latest poster or leaflet. When the material was ready for publication, helpers were appealed for. Andy Dingley, a computer buff, has provided desk top publishing assistance, also Fraser Robinson and Ray Weedon. Often this work took many hours of their valuable time.

The work of chair, treasurer, membership secretary, fund-raiser, report writer and publicity producers was continuous. These are the services which kept the DBC, like any other voluntary group, in existence, and continue to do so.

Besides the endless debating and lobbying against nuclear power, described in other chapters, there were the events and activities which raised money. The Druridge Bay Campaign quickly developed a regular series of three annual events. These were the Beach Run in the spring, the September Fair at Druridge Bay, and the Christmas Fair at Morpeth Town Hall.

The first of our three annual events, the Beach Run, was organised by Ray Scott, who had managed the first one for Friends of the Earth. He and his family and friends organised the runners, their number cards with safety pins, the recording of their times, even the design of the route.

We always tried to find a VIP to start the races. One year, 58-year-old Jack Thompson MP agreed not only to give out the prizes but to do the three mile run as well. While in London, he did his training on Hampstead Heath, which was just outside his back door. He got so fit that he was even able to race after his fellow MPs. He said, "I will be chasing my colleagues for sponsorship, and probably catching them with all this training I'm doing." At the time he said, "My wife Margaret is encouraging me and disciplining me for the run. She has put me on a calorie-controlled diet, no cakes, puddings or potatoes, but I'm refusing to eat lettuce. I hate it."[2]

The fact that Jack was doing the Beach Run made excellent publicity for us, including a front-page colour picture on the *Weekly Courier*.

The following year, 1988, we had a line up of MPs at the Beach Run. Lord Glenamara was our guest of honour and he distributed the prizes. Gordon Steel had met him on our lobbying trip to the House of Commons in February 1987, and he had agreed to help us in any way he could. Also there were Alan Beith MP, Bob Clay MP for Sunderland North and Jack Thompson, who did not run that year.

Lord Glenamara pledged his support for the Druridge Bay Campaign, and promised to use his influence in the House of Lords in any way he could. He told the Journal, "I have always supported the Druridge Bay Campaign because it would be an absolute tragedy if a nuclear power station was built here on one of the most beautiful stretches of beach in the country."[3]

The Beach Run continues to be an annual event, on one of the North's most scenic yet arduous routes.

The tradition of holding a Christmas Fair in Morpeth Town Hall was started by Helen Steen and Libby Paxton in December 1987. They were both very busy mothers of small children, but with the help of other Morpeth members, they worked very hard and raised over £450. Helen had made many Christmas decorations out of twigs and pine cones. Other members baked, stitched and hauled bric-a-brac and children's clothes. Goods were collected for a raffle from local traders.

Libby Paxton told the newspapers why she helped organise the event. "There are two purposes. One is to raise funds for a very important campaign, and to offer information about it. The other is to have a pleasant Christmas event for the general public."[4]

The Sturdy Beggars provided continuous singing, with medieval songs and instruments. The band were to become regular supporters at our events in future, and donated their services.

Jack Thompson had opened the Fair. In the following years, he or Alan Beith usually attended, and once it was the Deputy Mayor and Mayoress of Morpeth, Cllr and Mrs Clive Temple.

The biggest annual event taking more of our energies than any other came to be called the "September Fair". Friends of the Earth's Gathering at Druridge Bay had been in September 1983, on Derek Storey's farm at Druridge. In 1986 one of our members Hazel Lennox from Durham encouraged the DBC to hold a similar event by offering to manage it. Once again, Mr Storey and his farm manager Mr Phil Scott kindly allowed us the use of his field next to the Cairn.

In the dramatic setting of Druridge Bay, next to the traditional Northumbrian farm buildings of Druridge hamlet, with the roar of breakers behind the dunes, people came and refreshed their minds as to why it was so important to save Druridge Bay. The event attracted local musicians and craftspeople. There were games for children, stalls, sideshows, plenty to eat, and real ale. The weather for the 1986 event was on our side. It was bright though chilly. It was a great contrast to the storms and gale force winds of 1983.

Morpeth Footpaths Society supported the event by organising a walk round the potential nuclear power stations site which had been purchased by the CEGB. The DBC produced a leaflet called "A Druridge Walk", which described the natural history on the route, and included a section on the 14th century Chibburn Preceptory. This was a "hospital", a resting place for pilgrims travelling from Durham to Holy Island. A band led the walkers for the first stretch of their route.

The Fair has been held continuously since 1986. In 1987, Jane Gifford took on the task of organiser from her base in Newcastle. A range of people became responsible for a host of jobs.

Gordon and Joan Hanson and their family took on the "Bake-a-Cake for Druridge" stall. Joan Hanson's mother, 78 year old Jane Steel, baked a

The Stumbling Band from Newcastle lead the walk round the CEGB purchased site at Druridge Bay. Part of the September Fair 1986, this walk was led by Tony Shirley of Morpeth Footpaths Society. *Photo Peter Berry.*

Everyone Pulling for Druridge. Publicity organised for the September Fair, 1987. **Left to right,** Paddington Bear, Jack Thompson MP, Wansbeck Council Chairman Cllr Mrs Doris Brown and Helen Steen. Children include Wendy Scott age 4, Damien Thompson age 4 and Laura Gubbins age 7. *Photo Peter Berry.*

sumptuous selection of chocolate and coffee sandwich cakes, fruit cakes, pies and tarts. She was still doing this at the age of 83, in 1990. Nora Cuthbert of Morpeth and Sue Patience of Alnwick allowed their pictures to be taken for local newspapers of themselves in their kitchens Baking-a-Cake for Druridge, a simple but effective message to the public.

One year Wendy Scott organised a gymkhana in a corner of the field. Another year, kite makers from Newcastle organised children into making kites which could be flown on the spot. There has been a series of face painters, mask makers and general providers of children's games and entertainments.

The DBC's Support Groups took care of stalls such as bric-a-brac, books, plants and home produce. Teams of sandwich makers and tea brewers have taken their turns over the years. These include Ruth Simpkins, a student from Durham, Frances Lowes, a secretary from Newcastle, and Alan Priest, barman from Newbiggin. Ray Lashley, a miner from Ashington, lent us his caravan. It was not easy to cater for large numbers without water or electricity supplies, yet somehow it has always been done.

Almost every service has been donated. An artist member designed the posters. Northumbria Buses have provided the bouncy castle, hats and balloons. Castle

Gordon Steel and Druridge Bay Campaigners sweep the beach clean in readiness for the 1988 September Fair. Campaigners include the Sendall family, Jonatha and Cathy Robinson, Jane Gifford, Bridget Gubbins, Betty Greenwell and Spot.

Morpeth Borough Council provided the chairs and tables. Even the marquees were provided by North Tyneside Borough Council, until they could no longer help us as they were rate-capped. As well as providing the marquees, North Tyneside even sent along a team of workers to set them up. This saved us struggling with ropes, poles and canvas in brisk winds. Fortunately, since the gales of 1983, the weather has been reasonably kind.

The biggest annual hitch to overcome in organising the fair has always been provision of toilet facilities. Anyone who has been to Druridge Bay knows that there is nowhere north of Cresswell except the sand dunes for those in need. When several hundred people are gathered together, sand dunes are hardly the answer.

Ron Major remembered back to 1986. "At an early Fair planning meeting, we realised that we had set a date for which the Wansbeck District Council's mobile toilet block was not available, so I volunteered to try to obtain a unit from somewhere else. After a week or two of ringing around, the situation began to appear desperate. Any local authority who had mobile toilets had already booked them out. All of the commercial hire companies quoted me prices which would have taken between 30% and 50% of our budget. As the Fair day loomed nearer, a solution arose. I would build the Gents toilet myself.

"Therefore, the Friday prior to the Fair found me digging channels for the toilet in a corner of the field. After an hour or so I completed the task, feeling complacently satisfied with my efforts.

"However when I returned on the Saturday to erect the toilet, my complacency evaporated instantly when I realised I had overlooked a vital element in my calculations – the wind! As I struggled to complete the structure of struts, string and plastic, using over 5000 staples, a grim foreboding began to darken my mind. If the wind grew stronger, it could collapse, and someone would be prosecuted for indecent exposure.

"When I returned on the Sunday, my worst fears were confirmed. The wind **was** stronger, and the structure was flapping and billowing ominously. Throughout the day I kept suffering bouts of absolute panic. However, mercifully the structure held, I did not faint with fear, and no-one got prosecuted."

That year, we borrowed the Ladies toilet from a local Scout group. It was a primitive bucket with a seat, and not at all appreciated by the squeamish.

After 1986, the first thing we did when arranging the Fair was to book Wansbeck District Council's toilet block, and set the date round that. It always arrived and was greatly appreciated, until 1990, when it didn't. There was panic once again, as 12 o'clock opening time came around, and no toilet block had come. The Fair had to go on without it. And this year, unlike with an administrative hiccup the previous year, the Real Ale tent had come. One year there were toilets but no Real Ale, and the next Real Ale but no toilets... .

Harry McQuillan, of Amble DBC Support Group, has worked with his friends at supervising the toilet blocks on the occasions when they did arrive. He has helped with the car parking, along with Gordon Steel, another onerous but necessary task. Harry said, "It takes prime movers like Jane Gifford and Gordon Steel to get the Fair going. I have helped with toilets and car parking, and clearing up after the event. It was all good fun, and helped me realise how many people really **work** for the Campaign."

The overall management of the event was in Jane Gifford's hands. Accomplishing this during the summer months was quite a challenge, particularly in August when so many people are on holiday. In 1990, Jane was taken seriously ill, and forced to hand over the job. A team of emergency helpers took over, under the efficient steering of Wendy Scott from the office, and the Fair took place once again.

Mayor and Mayoress of Castle Morpeth Borough Council, Cllr and Mrs Ian Hunter, with Postman Pat and his Cat, and Jack Thompson MP. September Fair 1988.

The Druridge Bay Beach Bunny with Catherine and Rosemary Hall, Bridget Gubbins, Deputy Mayor and Mayoress Cllr and Mrs Clive Temple, and Sophie and George Paxton. September Fair 1989. *Photo Morpeth Herald.*

As our largest fund raising activity, the Fair has raised between £1200 and £1600 annually. It is important both for the money, and for the fact that it gets people out to Druridge Bay together, reminding us of the reason for the DBC's existence. It brings together members on the fringes of the organisation, who work hard on that day if on no other, and other members who come along merely to support it. Each year, more members of the public are attracted, as it becomes a regular event in the calendar of outdoor events in Northumberland.

• • •

Sometimes, a member comes up with a brainwave for an event which is quite distinct from our regular ones. Michael Kirkup, local writer from Ashington, was one such person. Michael wrote a poem about Druridge Bay, and then thought it would be possible to assemble an illustrated anthology of poetry about Druridge Bay and related themes.

He was extraordinarily successful in this. It was a long project, taking six months from the inception of the idea to the production of the book.

Michael knew many local writers, poets, illustrators and photographers. He wrote to them, asking for contributions, and also approached national figures.

The list of contributors is impressive, with local people rubbing shoulders with others of established stature. The foreword was written by Edward Bond. R S Thomas' poem "Tide Line" was adopted for the name of the book. Other poetry was by Mary Preston, Gerald Peel, Neil Taylor, Jon Silkin, Fleur Adcock, Mike Kirkup, Vernon Scannell, U A Fanthorpe, Anne Stevenson, Gene Groves, Jean Baker, Bruce Kent, Kath O'Brien, Maureen Hughes, Keith Armstrong, Linda France, John Earl, Dannie Abse, Peter Bennet, Vincenza Holland, Richard Adams, Hilda Helling, W J Findlay, Inabel Vallans, Roger McDonald, Edward Bond, Carol Dixon and Fiona Hall.

Photographs and drawings were by Birtley Aris, Julia Colquitt, Ken Ferguson, Phil Lockey, Robert Olley, Peter Berry, John Davis, Tom Fleming, Isabella Jedrzejczyk, Peter Lang, James Mellon, Ron Staines, Andrew Turner.

The following samples, selected by Michael Kirkup, give an idea of the range of issues covered in the book.

Edward Bond, the highly acclaimed playwright, wrote, "...and when the plans for the Druridge Bay nuclear plant are dust, the sands and rocks of Northumberland will still be evidence of our need for beauty, and the children playing in the clean sea will show our respect for human simplicity and the care we have taken of our communities".

U A Fanthorpe picked up the threads of an old woman's memories:
    Lots of us went, car-loads,
    Deebanks, Crawshaws, Mother, all of us.
    Children in troupes. Someone took a picture.
    And Mother there in black (*You look like the teacher*).

New Zealand-born Fleur Adcock kept the issue on a personal basis with her poem describing what happened to a girl on a working holiday at a Cumbrian farm near Windscale:
    Her family saw no cause for alarm –
    how could it do her any harm
    working there in the countryside?
    It would help to build her up, they said.
    But secretly broke her down instead,
                until she died.

Many poets wrote of the inevitable options to be faced, including Welshman, R S Thomas, in his poem "Tide Line":
    The waves come to the shore –
    where have the fish gone?
    A child bathed in the scaled waves –
    where has the child gone?

John Earl highlighted the peril:
    The marshes are wet with ghosts...
    In memory of Chernobyl...

Bruce Kent, in his new Psalm 151, deplored man's intrusion upon nature:
> What is man that you keep him in mind
> Nuclear man who ruins your creation

Gene Groves demanded a general awareness by the public, warning us not to be...
> Lulled by the soft sounds
> of the nursery rhyming sea.
> Deaf to the intrigues
> of the CEGB.

This theme was taken up by Fiona Hall who saw apathy as an ever-constant threat in her poem, "To an absent friend"
> Where were **you**?
> We all had great fun at the Fair,
> the Hungry for Change and the Peace Groups were there.
> Where **were** you?

Peter Bennet saw the problem as being one of human greed:
> Until we learn to do with less,
> and trust the dark,
> we're in on the conspiracy.

Jean Baker had no doubt that the issue was political:
> Minister – I've found the place,
> North-east England, miles away;
> I had a look last Saturday,
> You know what's there? Just open space!

Some poets focussed on the aftermath of a disaster similar to Chernobyl, including Anne Stevenson:
> And even then,
> there may be a language in which
> memory will be called 'letting in the sorrow'.

And Vincenza Holland's "Border Ballad" used the frightened voice of a child:
> Why do you hold me so tight mother;
> Why do your tears run so fast?
>
> Hush – very soon you will know, my son.
> Better not ask.

Yorkshireman Vernon Scannell lamented the loss of the everyday things:
> Those human things,
> The steaming amber streaming from the spout,
> The little rites and festivals,
> Music, books, and daybreak's beads of sound,
> Work's gruff and friendly challenge, its rewards,
> All gone as if interred in that dead ground.

As well as the poetry, the artwork was magnificent: Robert Olley produced a stark interpretation of "Border Ballad"; Ken Ferguson's "Sands of Time" was a vivid reminder of what might be; Birtley Aris favoured us with three striking interpretations; Phil Lockey, produced a stylised version of "Construction" and other poems; Peter Lang, a poignant version of "Mother in black, crying".

For photographs, Peter Berry from the Morpeth Herald took photos specially to Michael's requirements; two design students from Northumberland College of Arts and Technology, Andrew Turner and James Mellon, produced photos to match Jean Baker's and Jon Silkin's poems; Ron Staines' photo of Druridge Bay matched R S Thomas' poem Tide Line.

Professional photographers John Davies and Isabella Jedrzejczyk, who had produced an exhibition on Druridge Bay at Newcastle's Side Gallery, donated some of their work.

A few people to whom Michael wrote were unable to oblige for various reasons.

Lady Mary Wilson apologised: "I just can't write to order, but support your cause."

Mike Neville said, "I have my views on the situation, but my position as newsreader forbids me from airing them."

Playwright Alan Bennett observed, "I am hopeless at this sort of thing, but if I have an inspiration I will be in touch."

In June 1988, all the contributors to *Tide Lines* were invited to Morpeth Town Hall to help launch the book. It was a marvellous day. Twenty of the forty contributors attended for a buffet lunch, and to meet the press and TV. Much time was spent posing for newspaper photographers, and being organised for BBC Look North and Tyne Tees Northern Life camera crews. There was a poetry reading session in the afternoon. The event had an atmosphere of friendliness and good humour, and was highly charged with the energy of talented, dedicated people who were prepared to work for Druridge Bay.

Later, we all went home to watch ourselves on television. Both local channels covered it in depth. Tyne Tees concentrated on interviews with poets. For BBC, Mike Neville introduced the Look North programme. Michael Kirkup had already been taken out to Druridge Bay, and pre-recorded on the beach talking about his book. U A Fanthorpe's poem, "Druridge Bay in the 30s: the old lady remembers as she falls asleep", was read aloud, with sweeping shots of the bay in the background, and nostalgic photographs. It was wonderful for us to see our efforts being communicated across the North East like that. At the end, Mike Neville held up a copy of the book, saying, "*Tide Lines*, published by the Druridge Bay Campaign, £1.95." We couldn't have done better if we had spent thousands of pounds on advertising.

Over the next few days, photos and stories appeared in the local papers. There

Michael Kirkup, with authors and illustrators of *Tide Lines*, June 1988, in Morpeth Town Hall. Those present include: *Back row,* Ron Staines, Birtley Aris, Keith Armstrong, John Earl, Richard Adams, Peter Bennet, Fiona Hall, Jean Baker, Ken Ferguson, Roger McDonald: *Middle row,* Tom Fleming, Carol Dixon, Jim Findlay, Linda France: *Front row,* Bridget Gubbins, Jon Silkin, Mary Preston, Hilda Helling, Vincenza Holland, Gene Groves *Photo Peter Berry*

was even a photo, in the *Daily Telegraph,* of Jon Silkin reading poetry on the dunes of Druridge Bay.

Michael Kirkup had chosen a rather random selection of national, and even international, celebrities for his book, producing some unusual responses. He had written to the Pope. In reply, he was sent a copy of one of the Pope's speeches which he was given permission to use. He used the following selection:

"I feel inspired to say this to you... surely the time has come for our society to realise that the future of humanity depends, as never before, on our collective moral choice... our future on this planet exposed as it is to nuclear annihilation depends upon one single factor: humanity must make a moral about-face... Ladies and gentlement, it is for you now to take up this noble challenge."

Kingsley Amis was one of the very few who refused. He wrote, "I suppose that if I had ever heard of Druridge Bay, I was vaguely in favour of nuclear development there, whatever you may mean by that. Now that I have seen your list of the people you are asking for support, I am inclined to favour such development rather more strongly than before. (I except Pope John Paul II from the implications of the above.)"

Tony Jones, in his Midweek column in the *Journal*, made great play with the Kingsley Amis letter, suggesting that "maybe Kingsley Amis bridled when he saw the name of little-known poet Bruce Kent among the list of those who had been approached."[5]

Tide Lines was wonderful publicity, smoothly and successfully operated – or so it may seem. However, it had almost been an unprecedented disaster.

Mike recalled, "The first warning bell happened four days before the official launch event in Morpeth. I had been filming for BBC Look North, with Vicky Hornsby, out at Druridge Bay. During the shooting, while I was sitting on the dunes, I was asked to hold a copy of the book and refer to various passages. It was then that a page from the book took off in the wind, and disappeared among the dunes. I thought that that must have been a duff copy, and expelled the incident from my mind. Then, during the launch itself with everyone assembled in the Town Hall, media people, dignitaries etc., people began to bring copies back to the sales table to be exchanged."

He told me, "Somehow we stumbled through the launch, but the next day – when I also became a father for the third time – I went through half a dozen boxes of the books which were stacked underneath my stairs. Horrified, I found that every book in every box had been incorrectly bound. They literally fell apart when subjected to the least bit of pressure.

"To make matters worse, by then I had distributed vast numbers of copies in bookshops throughout the region from Alnwick to Durham. I rang you, and I think my first words were, 'Bridget, we are going to have to call in every copy of the book, and remove them from all the bookshelves'. Naturally, you were shell-shocked, and wondered if I wasn't overreacting. But after I explained the situation you agreed that this was the only course of action.

"But first of all I had to get in touch with the printer and demand that he reprint another issue of 2,000 copies, immediately, and without charge. He agreed to have a fresh batch delivered to me within a fortnight. It was the best I could get.

"In the meantime, I had to go to Newcastle, where many of the books had been placed, and tell retailers of the fault. Most were sympathetic, and said they would continue selling while offering their customers a swap when the new batch became available. And so a potential disaster was averted."

Despite that little drama, the book was very successful, particularly for a poetry book. It was immediately taken up by educational establishments. Matlock Further Education College in Derbyshire took a set of twenty. Various students approached us with the idea of of studying *Tide Lines* for their theses. Durham Johnston Comprehensive school visited Druridge, and encompassed the book within their GCSE studies. More recently, Cramlington High, Blyth Ridley High Bedlington High and King Edward VI in Morpeth have all taken full sets for teaching purposes, and Merton College, Oxford sent for more copies for their library.

Producing the anthology was important to Michael. He concluded, "Being editor of *Tide Lines* gave me a great deal of personal satisfaction. Since then, I have been lucky enough to be asked to write a couple of books and a musical, but I will always regard the Druridge Bay anthology as one of the highlights of my life."

As far as the Druridge Bay Campaign is concerned, the book brought in yet another sector of concerned people, enlisting us more support than ever, and sharing our message widely with the general public.

While *Tide Lines* was being produced, other work was going on. DBC Support Groups, loose groups of members from different locations, could be called on when events required it. As well as providing help at the events described already, they have been extremely useful at local political action. Their assistance during the process of Alnwick District Council's affiliation to the Druridge Bay Campaign is an example of this.

Alnwick District Council was one of the few immediately local councils which for many years had refused to oppose nuclear power at Druridge Bay. At a debate in August 1984, the majority clearly wished to take no stand. The minority, which called for a vote to re-think the issue, lost 7 to 16. .

In October 1984, the council decided to debate the motion that it should oppose the construction of a nuclear reactor at Druridge Bay. The vote after the debate split the council 12/12, and the motion was then defeated on the casting vote of Chairman Bill Mitchell.

Following this vote, in December Friends of the Earth took the petition to Alnwick. At a stall in the town centre, members prepared a large sign bearing the names of councillors and how they had voted. This created one or two heated conversations with councillors who had not wished to oppose nuclear Druridge, and who did not like seeing their names thus exposed.

Over the next few years, Alnwick's position did not budge. However, we felt it likely that the council was not reflecting the views of the electorate, and we continued to monitor the situation. By December 1987, when Councillor Hugh Philipson was Chairman of the council, the issue re-emerged. Councillor Albert Davidson and the late Councillor Geoff France had attended the recent DBC Jobs Forum, and argued convincingly that nuclear power at Druridge represented a real threat to local jobs. The council agreed to invite speakers from both the DBC and the CEGB. After hearing them, there would be a vote on whether or not to affiliate.

The meeting was arranged for June 1988. The months between December and June were available to be used by our members in Alnwick District to highlight the issue. We had two support groups in the area. There was one in Warkworth/Amble co-ordinated by Debbie Butcher, and one in Whittingham under Fiona Hall. Cath Young and Ken McDonald in Alnwick agreed to set up a new group there.

These members immediately started talking to their councillors and gathering signatures for a petition. Fiona Hall remembers a somewhat problematical interview with her Independent councillor, Mrs Marion Guiry.

She said, "We knew she wasn't on our side, but I went to see her anyway. It was a very difficult occasion. I sat in her kitchen and she really did tear a strip off me. She was very hostile towards people who came into Whittingham and Glanton to live, and wanted to do all this campaigning for something that was outside the villages instead of getting involved in issues in the villages themselves. I realised it was a difference in perspective. I felt that Druridge was part of our environment. It is only 25 miles away. We would be affected as much as anyone. There wouldn't be a community here if anything went wrong at Druridge. She mentioned digging out an old well in Glanton. Later John Griffiths (one of our DBC supporters) went to see her about this, and had an amicable meeting. But on this occasion, I left feeling like a wrung out lettuce."

Fiona put a petition in the village shop. It elicited 104 signatures, and from her personal knowledge of local people, she knew they were from a cross-section of political viewpoints.

Ken McDonald and Cath Young, with other members, agreed a programme of activities to focus on Alnwick councillors. They posted out information to them all. They set up a mobile barrow in the market place and round the town centre, on which they had a petition to Alnwick council, and organised a flag day.

Ken recounted his own feelings and the reaction of local people in this conservative Northumbrian market town. "It took some courage to actually get out and do this campaigning. But to our suprise, it worked. It was like democracy in action. We found we were a focus for people to express their concern. There were so many people who were willing to put their names on the petition, people who were not initiators of action, but who were deeply concerned about nuclear power at Druridge Bay. There were of course some who thought that nothing we could do would stop 'them' from building nuclear power stations if they were determined. And there were a very few who were in favour, usually because they thought they might get a job there rather than because they liked nuclear power particularly."

Cath remembered, "As well as local people, we met tourists from all over the world. I particularly remember the Americans, who were much more advanced in environmental campaigning than we were. They had been against nuclear power for over a decade, whereas we had just started."

Then Alnwick DBC Support Group placed a public notice in the *Northumberland Gazette*. It outlined reasons for opposition to nuclear power at Druridge Bay, and invited local people to contact the Chairman of Alnwick District Council with their views. This action had a surprising effect.

One day, Ken McDonald had a telephone call from the irate Chairman of Alnwick District Council. Councillor Hugh Philipson complained bitterly that Ken and his friends had placed the public notice in the newspaper, without specifying that it was put in by Alnwick DBC Support Group. The notice asked people to send their views to the Chairman of Alnwick District Council. The Chairman had concluded, with some reason, that it looked as though he had placed the notice in the newspaper, and was soliciting people's views. The implication was that the body who had inserted the notice was opposed to nuclear power. As Councillor Hugh Philipson was one of the most ardent opponents of the Druridge Bay Campaign, he was furious.

Ken put the matter right as well as he could by sending a letter to the *Northumberland Gazette* the following week, explaining the error. Tongue in cheek, he wrote, "We invited readers to make their views known through the office of chairman, rather than to any individual, as that office fulfils the function of impartial representative of Alnwick District Council as a whole."

That incident kept the issue boiling locally. Meanwhile, signatures were being gathered. Warkworth DBC Support Group made contact with their mainly friendly councillors, and time for the big debate arrived.

The speakers, for and against nuclear power at Druridge, addressed the council in June. Cllr Janet Cann of Castle Morpeth and I spoke for the DBC, and our old friend George Johnston for the CEGB. It was one of the most challenging situations I had ever found myself in. DBC members thronged the gallery, offering their moral support.

One month later, Alnwick District Council, having deliberated to its satisfaction, voted 18–8 in favour of affiliation to the Druridge Bay Campaign. We were jubilant. It was a complete turnaround from 1984. Now we could really say that all local authorities round Druridge Bay supported us.

• • •

The people who support us vary from the most ordinary member of the public – if there is such a thing – to the most renowned. We were able to cultivate a list of famous people who lent their names to our cause. At the time of the Sizewell Inquiry, Friends of the Earth in London produced an advertisement for the national papers with a long list of several hundred VIPs who opposed the building of the Sizewell nuclear power station.

We took the list to Who's Who, found addresses and wrote to as many of these people as possible, asking them if they would support our campaign to fight nuclear power at Druridge Bay too. Many agreed to do so, as can be seen in the list below.

Tide Lines contributors added to our now growing list of VIPs. Another sector was that of our supporters in the political arena.

Alan Beith and Jack Thompson MPs had been our stalwart supporters since 1983. Gradually, we obtained the support of more and more local MPs. After our lobby to the House of Commons on the Sizewell report, virtually every North East MP supported us.

Here are some extracts from their correspondence.

Nick Brown MP Newcastle upon Tyne East, 24 January 1984: "I am opposed to the construction of a nuclear power station at Druridge Bay. I support what I understand to be the present position of the General and Municipal Workers' Union which is that Great Britain should not construct any further nuclear power stations."

Dr David Clark MP South Shields, 1 April 1985: "It is insulting to build a nuclear power station on top of the Northumberland Coalfield. Coal needs to be the mainstay of our energy policy. In addition it is wrong to site a nuclear power station amidst such beautiful countryside."

Bob Clay MP Sunderland North was a long-standing supporter. We tried to encourage him to participate in the Beach Run in 1988. He wrote, "I might even have a try at the three mile course, but quite honestly would want to know who else was participating (I don't want to look a complete wally). If there are any other MPs participating who are as unfit and knackered as myself, I don't mind joining in." Poor Bob. No other MP would join him, so he settled for attending as a supporter.

Dave Clelland MP Tyne Bridge, 18 February 1987: "I would be delighted to become a patron of the Druridge Bay Campaign, and to help the Campaign in any way I can."

John Cummings MP Easington, 28 June 1987: "I am very interested in the work of the Druridge Bay Campaign, and I would be happy to be associated with it by becoming a Patron."

John McWilliam MP Blaydon, 6 February 1985: "I fully support the objectives of the Druridge Bay Campaign. The future power needs of the North would be far better met by Combined Heat and Power Schemes, and not by destroying an area of outstanding natural beauty."

Marjorie Mowlam MP Redcar, March 1988: "It is important to keep Chernobyl at the forefront of people's minds. The effects are still being felt in the UK and in the rest of the world. The nuclear industry has learnt no lessons from Chernobyl."

Joyce Quin MP Gateshead East, 13 May 1987: "I will certainly oppose the construction of any power station at Druridge Bay. I am fully convinced that this beautiful stretch of our coastline should be preserved for future generations to enjoy."

Other MPs who sent messages of support include Hilary Armstrong Durham NW, Ronnie Campbell Blyth, Frank Cook Stockton North, Jim Cousins Newcastle Central, Donald Dixon Jarrow, Derek Foster Bishop Auckland, Doug Henderson Newcastle North, Chris Mullin Sunderland South and Gerry Steinberg Durham City.

Only a few local MPs were unsure about giving us their support. By 1987 there were only two Conservative MPs between Scotland and County Durham. One, Neville Trotter Tynemouth, would not go quite so far as to agree to be on our VIP list, but wrote, "I shall continue to oppose the construction of a power station at Druridge Bay and press for a coal-fired station to be built elsewhere in the region."

Alan Amos Hexham was the other. Although friendly, he would not take a firm stand opposing nuclear power at Druridge. He never gave us anything more meaningful than comments like the following: "I can assure you that I am continuing to take a keen interest in all aspects of nuclear power as I realize how important this matter is to my constituents and Northumberland."

Ted Garrett Wallsend (Labour) shifted his views over the years. In January 1987, he wrote to Gary Craig: "I do not share your opinion that there would be total opposition to nuclear power as a means of energy. I believe that nuclear power should be a part of the broad energy policy for the country." By May 1987, he had written to Councillor Ashbridge, "I have already stated my opposition to the construction of a power station at Druridge Bay."

The doubters on the Druridge issue were few. After our lobbying in the House of Commons and the House of Lords, described in the next chapter, we were able to add many more influential names to our VIP list.

By July 1989, our VIP list was as follows:

Artists, Writers, Academics, Poets, Actors, Scientists, Musicians etc.

Dannie Abse
Fleur Adcock
Birtley Aris
Keith Armstrong
Jane Asher
Pat Arrowsmith
Beryl Bainbridge
Alan Bennett
Peter Bennet
Edward Bond
Prof Gustav Born
Richard Briers
Prof D Bryce-Smith
Julie Christie
Sinead Cusack
Rev A H Dammers
Douglas Dunn
John H Earl
U A Fanthorpe
Dame Elisabeth Frink

Edward Fox
David Gentleman
Brigadier Michael
  Harbottle
Prof H Himmelweit
Vincenza Holland
Prof Dorothy Hodgkin
Bruce Kent
Laurie Lee
D & E Leggett
Sir Charles Mackerras
Richard Mabey
Roger McGough
George Melly
Christopher Milne
Dr Paul Noone
Bill Oddie
Robert Olley
Nigel Planer
Jonathan Pryce

Diana M Quick
Dame Diana Reader
  Harris
Lord Ritchie
Richard Rogers
Dr Wendy Savage
Vernon Scannell
Jon Silkin
Sir K and Lady P
  Spencer
Anne Stevenson
Juliet Stevenson
Rosemary Sutcliffe
Prof R L M Synge
R S Thomas
Prof N Tinbergen
John Williams
Michael Williams
Shirley Williams

Local MPs, MEPs & Lords

Gordon Adam	Dave Clelland	Doug Henderson
Hilary Armstrong	Frank Cook	John McWilliam
Alan Beith	Jim Cousins	Lord Morpeth
Tony Blair	John Cummings	Marjorie Mowlam
Nick Brown	Donald Dixon	Chris Mullin
Ronnie Campbell	Lord Dormand	Joyce Quin
David Clark	Derek Foster	Gerry Steinberg
Bob Clay	Lord Glenamara	Jack Thompson

Other MPs, MEPs, & Lords

David Alton	Andrew Faulds	Stan Orme
Lord Beaumont	Lord Hatch	Lord Ross
Tony Benn	Lord Hylton	David Steel
Robert Brown	Lord Irving	Gavin Strang
Malcolm Bruce	Lord Jenkins	Lord Williams
Anne Clywd	Joan Maynard	Lord Winstanley
Lord Ennals	Michael Meadowcroft	

Just as we need the support of celebrities, so we could not do without the workers who do the routine chores. There are people who have come in on a regular basis to the office, helping with the tedious jobs, including Sally Dawson from Newbiggin, Alison Hopper from Morpeth and Marianne Peer from Longhorsley. There have been people who have offered hospitality when we have had guests to look after, like Nora Cuthbert of Morpeth and Jane Gifford and her friends in Newcastle. When we have needed artistic skills, Julia Hilton, a printer, produced a glorious banner, and together with Jane Gifford designed and produced our T-shirts for children which say "Build a sandcastle at Druridge Bay". When we have needed photographs, first Ian Barkley from Acklington and then Gordon Hanson from Gateshead made themselves available, often at short notice. When we needed new Druridge mugs after Hilary Town left the area, Irene Ismay from Ulgham provided a new design. Richard Stourac, drama lecturer at Newcastle Poly, enlisted the support of his students in a series of plays and sketches about the nuclear industry and Druridge Bay. When there was a call for someone to appear on TV in a religious chat show, we were able to call on Carol Dixon of Pegswood to fill the slot.

Over the course of many years, campaigners come and go. People can get weary from sustaining the effort of helping protect Druridge Bay, especially when the threat is not immediately imminent. Sometimes we as organisers would get dejected at low attendance at meetings, and be forced to re-assess what it is possible to achieve, just like any other struggling voluntary organisation. The nuclear industry by comparison can pay its workers well, and afford propaganda to maintain its image. Fortunately our councils who supplement our office and worker expenses enable us to maintain a basic level of activity. Thus we can keep members informed and involved.

The Druridge Bay Campaign is its members and supporters. Without every one of the people mentioned in this chapter and the many others who are not, it would not have kept going for over a decade. From Ron Major digging his earthen toilets in the wind at the Druridge Fair to Lord Glenamara speaking for us in the House of Lords, it is **people** who are defending Druridge Bay. Every one is important.

NOTES
1. *Journal*, 5 10 84
2. *Journal*, 6 4 87
3. *Journal*, 3 5 88
4. *Morpeth Herald*, 10 12 87
5. *Journal*, 2 6 88

9. PRESSURE BUILDING UP
1988 – 1989

In 1988, the Central Electricity Generating Board was still in command of a mighty industry, with a total monopoly of electricity generation and supply, and a determination to expand nuclear power. It was protected by an invincible wall. Government support for its plans and its own enormous financial power rendered it seemingly indifferent to the attacks of its critics. Yet during 1988 and 1989 its foundations were being undermined and cracks were forming in its facade. At first these changes were scarcely visible. Their consequences were not fully forseen by those most affected, both within and without the nuclear industry.

During this period, there were two major developments. One was the forthcoming privatisation of electricity, and the other was the publicity given to global warming, better known as the Greenhouse Effect.

The questions and answers, challenges and arguments, and coverage in the media throughout these two years were intense. All at once, it seemed that there was a new public awareness of the environmental consequences of energy use. Everyone was asking questions about burning fossil fuels in huge coal-fired power stations. Some were suggesting that nuclear power was the clean source of electricity for the future. Previous environmental concerns about acid rain from coal-fired power stations faded by comparison, and the consequences of nuclear accidents such as had happened at Chernobyl seemed to be forgotten.

With electricity privatisation, there was a sudden interest in the economics of nuclear power. Was it really as economic as had been supposedly demonstrated at Sizewell?

Meanwhile the work of the Druridge Bay Campaign went on. The electricity industry was to be privatised. The CEGB was to be broken up into independent components, consisting of two generating companies, a grid company and regional electricity supply companies. Nuclear power was to be in the hands of National Power, which was to have 70% of the CEGB power stations including the land at Druridge Bay. This meant that a private company would be the target of our efforts. It would be the shareholders, not the government, that would have to be lobbied or persuaded that new nuclear power stations were not a good idea, especially on an unspoilt coastline.

The Privatisation of Electricity White Paper was published in February 1988, and the DBC debated its response. It was decided that the arguments for or against privatisation of electricity were not within our remit, as we were a campaign against nuclear power on a specific site. However, we were very interested in the effect that electricity privatisation would have on new nuclear power station development, and decided to lobby for our interests.

What effect could the Druridge Bay Campaign's efforts in lobbying Parliament have on the Electricity Bill? Although well established in the North East, the DBC could have only a most minor influence in national affairs. Mrs Thatcher's government, heavily inclined towards both privatisation and nuclear power, and with its large majority in the House of Commons, would be able to push through its desired legislation with ease. However the privatisation of electricity provided us with the opportunity to make our opposition to nuclear power at Druridge Bay heard. We were determined to make the most of it.

In this work, Adrian Smith, Newcastle City Council's representative on the Steering Committee, became greatly involved. We agreed that there were three points on which the Druridge Bay Campaign should lobby during the passage of the Electricity Bill through the Houses of Parliament.

They were as follows. First, the bill should not protect nuclear power from market forces. Favouring of nuclear power would encourage more nuclear power stations to be built, and increase the threat to Druridge Bay.

Second, we sought a guarantee of a full public inquiry in the event of a planning application being made at Druridge Bay. One of the consequences of protecting nuclear power by law would be that the full scope of a public inquiry would be reduced. We wanted to be certain that all the economic, environmental and social consequences would be taken into account.

Third, we sought much improved legislation on energy efficiency. Electricity generation makes a major contribution to environmental pollution. Greater efficiency in the generation and use of electricity would mean less nuclear waste, acid rain and greenhouse gas emissions. It could also mean fewer new power stations.

Such were the arguments, backed up with an impressive file of reports and information, which we compiled ready to lobby the Houses of Parliament.[1]

We had briefing sessions with both Jack Thompson and Alan Beith on how best to present our arguments. We were advised to lobby for support from local MPs in the North East, rather than in London. They told us that whereas going to London makes the headlines, the real work is done in the constituencies.

Alan Beith agreed to book a room for us in Westminster where we could meet local MPs. They told us about the second reading debate in the House of Commons, followed by the standing committee which considered the bill clause by clause, and where amendments could be put forward. They both agreed they would try to get on the standing committee, although with its large majority of government MPs, it was unlikely to be influenced. After the House of Commons stage, the Bill would then go to the House of Lords.

It was of invaluable assistance to have the support of these MPs. Not only were they able to guide us on procedural matters, we knew they would speak up for

Druridge Bay in the House of Commons on every opportunity.

We planned a programme of action, and set to work to enlist the support of our members. Fortunately, for this year of great activity, the DBC was able to hire a clerical assistant. Joan Watson, from Shilbottle near Alnwick, took care of the numerous letters and messages that passed to and fro.

All our members were encouraged to write to their MPs. We particularly targetted Neville Trotter and Alan Amos, the only two local Conservative MPs, who as members of the party in government were in a particularly good position to speak out on our behalf.

Adrian Smith produced a briefing pack summarising our three main arguments, and a local firm called FormWord designed and produced a green glossy cover for us free of charge.

In November 1988, a coach-load of members went to the House of Commons in conjunction with the national lobby organised by Friends of the Earth. The FoE lobbyists were campaigning particularly against protection for nuclear power. Our points were focussed on Druridge Bay but our aims were broadly similar.

We were met by BBC and ITV, giving our case an airing to the electorate back home in the North East. Alan Beith and Wansbeck Councillor George Ferrigon were both interviewed by local TV.

Druridge Bay Campaigners at the House of Commons, lobbying on aspects of the Privatisation of Electricity Bill, Nov 9 1988. Barbara Devereux and Toni Stephenson hold the banner. *Photo Gordon Hanson.*

After the media attentions, Alan Beith took us to the Committee Room, bypassing the long queue of Friends of the Earth campaigners from across the country. We were joined inside by Marjorie Mowlam MP for Redcar, Joyce Quin MP for Gateshead East, Gerry Steinberg MP City of Durham, Ronnie Campbell MP Blyth and David Clark MP South Shields. Unfortunately Jack Thompson's wife was ill and he could not be with us. Several other MPs sent messages of support. Adrian Smith presented our case for no protection for nuclear power, a guaranteed full public inquiry and regulation for energy efficiency.

Formal events over, we met other campaigners from the groups opposing the PWRs around the country. In the *Red Lion* in Whitehall, we met people from Sizewell, Hinkley, Wylfa, Dungeness and Winfrith. For many members, it was the first opportunity they had had to compare experiences.

Mrs Paddy Davidson and her friend Susan Wilson from Cramlington came on the lobby, the first time for either of them. Mrs Davidson recalled, "We assembled at the House of Commons with our banners. The MPs came out to meet us. I felt a little bit embarrassed as I had never stood with a placard before. I was very impressed with Alan Beith. He was a real gentleman. Then Ronnie Campbell came out, wanting to know if we were the people from his constituency. He couldn't stay for our meeting, but arranged to meet us afterwards.

Inside a committee room in the House of Commons, Druridge Bay Campaigners give a presentation to MPs. Those pictured include Alan Beith MP, Bridget Gubbins, Barbara Devereux, Adrian Smith and Paddy Davidson. Nov 9 1988. *Photo Gordon Hanson.*

"We had our photos taken by local newspapers and then went into the House of Commons Committee Room. The MPs seemed to be really interested in what we were saying. I sat quietly. I wasn't overawed. I was enjoying it, though I didn't speak. I did feel the MPs wanted to rally round us.

"Later Ronnie Campbell took us to the bar and bought us some drinks. Other MPs came in, and a deputation of three miners from Northumberland joined us. After that we went to the House of Commons shop and bought a few souvenirs before having to leave for our bus.

"It was a day I remember very well. I felt we got our points across. Back home, my daughter saw it on TV. Going to London shows people and the government that we take Druridge Bay seriously. If they try to go ahead with plans for Druridge Bay, we will be ready for them."

As well as going to the House of Commons, an important part of our strategy was to lobby MPs locally. Using our new computer, and with Joan Watson's assistance, we were quickly able to write to all DBC members in the two constituencies of Neville Trotter and Alan Amos. We asked members if they would like to meet or write to these MPs, and many did. Those who wrote received replies, including word processed copies of the standard response from the Department of the Environment.

Three members went to visit Neville Trotter. They were John Hartwright and Jane Grimes from Whitley Bay, and Adrian Smith.

Adrian Smith said, "Neville Trotter was welcoming and warm towards the Druridge Bay Campaign. A concern for mining jobs came through very strongly, and he promised to do what he could to help. We felt that the meeting had been worthwhile and we had gained an ally."

At this meeting, he re-stated that he was totally against nuclear power stations at Druridge Bay, though not against nuclear power as such. He questioned Adrian closely on alternatives, particularly on gas-fired combined heat and power. He agreed to send our briefing pack to Cecil Parkinson, then Secretary of State for Energy, enclosing a letter with his comments. When the reply came from the Department of Energy, it was the same letter other members had already received signed by Michael Spicer, Under Secretary of State for Energy.

The meeting with Alan Amos had a different flavour. Five of his constituents went with Adrian Smith to Prudhoe Conservative Club. They were Maggie Hopkirk, a Citizens Advice Bureau worker, Thelma McGuckin, a nurse, Marilyn Kendall, a teacher, Neil Weatherly, a Stannington parish councillor and Birtley Aris, an artist.

Adrian recalled, "Alan Amos gave us a fair hearing. Unfortunately for us he favoured nuclear power in principle, and Druridge Bay was 25 miles from his constituency. We left feeling that we had put the message across, but without much likelihood of changing his mind."

Neil Weatherly took notes at the meeting. His criticisms were harsher. He reported that Alan Amos did not oppose nuclear power at Druridge Bay, and had no confidence in the role of renewable energy. Neil Weatherly said, "There is no doubt that if this government supports the siting of a nuclear power station at Druridge Bay, we will get no support from this MP whatsoever."

Every time we undertook one of these exercises, we would notify the papers, informing local people of our arguments and actions. Many of our members themselves were approaching their MPs for the first time. All our actions led to an increase in understanding of the issues and the decision-making process.

We continued to notify the media about the efforts of our MPs in the House of Commons. Three local MPs spoke in the second reading debate on the electricity bill. Alan Beith drew attention to the lack of proper regulation for energy efficiency, and objected to the market distortion which would result from protecting nuclear power. Jack Thompson specifically brought Druridge Bay into the debate, and pointed out that the lack of regulation would mean the continuing run down of the coal industry. Tony Blair, at that time the shadow spokesman for energy, was also a local MP from Sedgefield in County Durham. He argued fluently against the clauses which protected and subsidised nuclear power.

During January and February 1989, the House of Commons Standing Committee on the Privatisation of the Electricity Bill went through the bill clause by clause. Tony Blair was on the committee, and to our delight, Alan Beith had also obtained a seat. Giles Radice from Durham North was another member so the North East was well represented.

The government majority would not permit any changes of consequence. The standing committee debated for three and a half hours on the two brief energy efficiency clauses in the bill. Five improved amendments were put forward, two of them by our local MPs, Tony Blair and Alan Beith. However, the amendments were quashed by the government majority, 17 votes to 14.

As publicity officer, it was my job to obtain information about relevant debates and discussions. Our helpful MPs would send me Hansard, and I would then to turn the information into a press release or contribution to the letters page of local papers. It was a mutually supportive arrangement. We asked our MPs to promote our case, and then in return did our best to ensure that their support was fully recognised. Equally, we had no qualms about giving publicity to those MPs who did not support our case.

While the Electricity Bill was making its way through Parliament, the national debate on energy issues was taking place in parallel. In September 1988, the public inquiry into Britain's second PWR at Hinkley had started, where the CEGB admitted that nuclear power could be more expensive than coal.

Other evidence at the Hinkley Inquiry showed that wind turbines would be more economic than nuclear power stations, and that the cost of building Sizewell B had increased by 1 million, 7% more than expected.[2]

All these were shocking admissions, reducing the confidence of those in the private sector who might be buying the soon-to-be-privatised electricity industry, complete with its nuclear component.

Then in March 1989, Lord Marshall, chairman of the CEGB, closed down Britain's oldest Magnox nuclear power station at Berkeley in Gloucestershire. It took 10 seconds to do it, but as *Guardian* reporter Patrick Donovan commented, "it will take for all eternity to store the radioactive waste," and 100 years before the plant itself can be pulled to pieces.[3]

Once again, costs came into question. Who will pay for final decommissioning, of this and the CEGB's other 13 nuclear power stations? The government? Shareholders? Cost estimates ranged from the CEGB's £300 million per nuclear power station to John Large consulting engineers' £750 million.[4]

As well as costs of nuclear power, much publicity was being given during 1988 to the Greenhouse Effect. This was believed to be caused by excess burning of fossil fuels which give off carbon dioxide in the process. Too much carbon dioxide in the upper atmosphere was trapping heat, causing a rise in global temperature. Other gases contributed to global warming, but carbon dioxide was the greatest offender. Such was the theory gaining wide acceptance. Among the main contributors to the carbon dioxide emissions were the coal-fired power stations. This gave the nuclear industry a new burst of enthusiasm. At last nuclear power could claim to be the clean, green source of energy.

Campaigns like ours which had promoted coal as an acceptable alternative to nuclear power, as long as pollutants causing acid rain were removed, had to think again.

By September 1988, the *Daily Telegraph* was reporting, "Globally about three billion tonnes of carbon dioxide is added to the atmosphere each year, mostly from the burning of fossil fuels" and "Nuclear power plants produce no carbon dioxide. Environmentalist groups will have to examine more carefully their opposition to nuclear power."

Our earliest response to this was published in a letter to local papers in October 1988. "The answer to a nasty problem is not to substitute it for an equally nasty solution. Both coal burning and nuclear power for electricity have mammoth environmental consequences. Whereas coal burning is one of the contributing factors in causing the greenhouse effect, nuclear accidents have appalling consequences also, and the problem of disposing of nuclear waste, poisonous for centuries, is horrific."

We began to realise that the position we were taking in our lobby to the House of Commons also applied to the Greenhouse Effect. Energy efficiency, a key concept relating to energy and the environment, would both reduce the need for new power stations and lessen the Greenhouse Effect. It was a new idea for both us and the general public in the summer of 1989, though by now has become an everyday household phrase.

The TV programmes and newspaper debate continued, often emphasising the role of coal-fired power stations in causing the Greenhouse Effect. BNFL ran a full two-page series of articles in the quality newspapers during 1989, with the slogan "The Greenhouse Effect – We have the power to help prevent it".

It took up two full pages, one describing all the various contributors to the problem, full of smoke and dark skies, and one with a clean nuclear power station with no chimneys nestling among fields and trees, near a little thatched cottage. The nuclear plant was only a little taller than the trees, and there was a mere wisp of a pylon in the background. Its conclusion was, "Fortunately, we do already have a secure source of clean energy for the future. Nuclear power." There was no mention of radioactive waste, nuclear accidents, or decommissioned nuclear hulks.

Also jumping on the bandwagon, the CEGB calculated that ten new nuclear power stations could be commissioned by the year 2000 (one of which would certainly be at Druridge Bay) and estimated their contribution to reducing carbon dioxide emissions.[5]

One day, one of our most devoted members, Barbara Devereux from Morpeth, came into our office. She said, "You know Bridget, I hate the thought of nuclear power at Druridge Bay, but perhaps we will have to have it, because of the Greenhouse Effect". The realisation that the arguments of the nuclear industry were getting through to our own members had a great effect on me. We simply had to get to grips with this issue. The Steering Committee agreed, and it became my task to prepare an information leaflet on the subject.

While preparing this leaflet, one difficulty became clear. Our remit as a campaign was to oppose nuclear power at Druridge Bay. We had never officially changed from our original position not to be anti-nuclear. The nuclear industry was saying that nuclear power was an answer to the Greenhouse Effect. We had to say it was not. How could we do that without arguing against nuclear power? How could we say that nuclear power at Druridge Bay would make only the most minimal difference in reduction of greenhouse gases, therefore it should not be built, without laying us open to accusations of "not-in-my-back-yardism."

Equally, to argue that a coal-fired power station alternative would make very only a small additional contribution to the problem was dubious. This was one of the first examples of the widening of the energy and environment debate going on all around us, which affected our activities. It was hard for us to participate if we stuck closely to our narrow remit of opposition to nuclear power at Druridge Bay, which had served us well in the past. Getting the wording right in our Greenhouse Effect leaflet was problematical.

When ready it showed the contribution of coal in comparison with other causes such as oil burning for transport, and CFCs. It included hard information on the dangers of nuclear power, nuclear waste and decommissioning. It showed how nuclear power indirectly emitted carbon dioxide, and that this would radically increase with an expanded nuclear programme.[6] It offered solutions based on energy efficiency and renewable energy, cheaper and safer than nuclear power.

Every piece of information in the leaflet was carefully referenced, and a team of six scientists and six lay members checked every word. By the time it was finished, we had a substantial basis for our argument, and Jonathon Porritt who was at that time Director of Friends of the Earth, came to Newcastle to launch it.

Greenpeace at this time, July 1989, ran full-page advertisements in national newspapers to counter the claims of the nuclear industry that it had the answer to the Greenhouse Effect. "SCIENTIFICALLY SPEAKING, IT'S JUST A LOT OF HOT AIR," said the ad.

An impressive list of senior scientists from British universities, hospitals and other establishments signed the statement, "NUCLEAR POWER IS NOT AN ANSWER TO THE GREENHOUSE EFFECT."

We wrote to them, enclosing our new leaflet, and asking us to support our campaign. As a result, we added 10 new names to our list of VIPs. They were:

Dr Michael Whelan, University of Oxford
Prof David Smythe, University of Glasgow
Prof Robert Hill, Newcastle upon Tyne Polytechnic
Prof John Burland, Imperial College of Science
Dr Stephen Moorbath, University of Oxford
Dr Reginald Rainey, formerly Ministry Overseas Development
Prof Maurice Wilkins, Emeritus Prof, University of London
Prof Robert Hinde, University of Cambridge
Prof Keith Puttick, University of Surrey
Prof Brian Jarman, St Mary's Hospital, London

• • •

While we were following the standing committee proceedings in the House of Commons and getting to grips with arguments on the Greenhouse Effect, we were preparing for lobbying the House of Lords. This time, instead of having helpful MPs to advise us, we sought out the Bishop of Newcastle and Lord Glenamara.

One January afternoon in 1989, Adrian and I found ourselves having tea with the gentle, elderly Bishop Alex Graham in his old-fashioned sitting room in his house in Gosforth. His housekeeper brought us tea, and his much-petted golden retriever obligingly shared her sofa with us.

The Bishop was kind, and shared our concern for Druridge Bay. He seldom attended the House of Lords, so did not see that he could be actively involved. He recommended various local lords to whom he thought we might apply for support. These were Lord Allendale, Viscount Ridley, the Duke of Northumberland, Lord Rippon of Hexham, and in particular our friend Lord Glenamara. He suggested that we contact other bishops, especially those closer to London, who were more likely to attend the House of Lords. We realised that our campaigning methods were neither his style nor part of his role, but we were grateful for his interest and moral support.

We had prepared our new *Briefing Notes,* an up-dated version of our House of Commons pack. Each of the twenty-four bishops of England was sent a copy. We rang the local library to see how bishops are addressed. We learned that instead of writing "Dear Sir", or "Dear Bishop", we should put "My Lord". Laying aside any plebian demurring, we did this. In the letter, we asked them to support our case, and speak up for us in the House of Lords. We asked them if they could attend our meeting in Westminster, and if not for a note of support for our aims.

Ten bishops replied to our letter, some explaining that they were not able to play an active part in the House of Lords. Most had discreet, brief messages which tactfully conveyed understanding without expressing an opinion. The Bishop of Ely's was one such letter. His secretary wrote, "The Bishop has asked me to thank you for your letter together with a copy of the *Briefing Notes.* He has read what you say with interest, but unfortunately does not think he will be able to attend your meeting. He asks me to send you his best wishes."

The main benefits we got from this exercise were media attention and knowing that we had drawn Druridge Bay to the attention of the wider world.

Lord Glenamara was vital to our next stage of action. One Sunday afternoon, I went to meet him and his wife at Wallington Hall tea room. Over tea and scones, he advised me on procedures, and we planned a programme of action. He agreed to book a committee room for us to give a presentation at the House of Lords, and advised us on the timing for this in relation to the Electricity Bill debate. He told me that any amendments we wanted sympathetic lords to put forward should be exactly worded. This was an improvement from our House of Commons presentation, where we had described our case and left the formulating of amendments to MPs.

He explained that the committee stage, equivalent to the standing committee in the House of Commons, is done in the main debating chamber of the House of Lords. It is open to any interested peer to attend. At this stage, amendments such as we might suggest can be introduced, and there are many opportunities for divisions, ie voting. When this occurs, whips call peers in to vote.

He also explained that there is a large number of Cross Bench peers. If Cross Bench peers vote with Opposition Peers, they can command a majority. Thus it was important for us to gain their support if we could. He promised to send us a list of lords, both local and national, who might share our aims. He also told us that he would try to enlist support from the opposition spokesman for energy, Lord Williams of Elvel.

Having such first hand advice from such an experienced politician was absolutely invaluable. We now knew what to do.

Once back in the DBC office, a fleet of letters was sent on its way to lords both local and national, to our members encouraging them to write to lords and to join us on the lobby, and to the press informing them of what we were doing. It was fun collecting the mail each morning, as the smart cream and red envelopes from the House of Lords came in with replies.

It was interesting to note that there was by no means such a high rate of response from lords as there would have been from MPs in a similar action, and the tone was different. Lord Ridley of Blagdon, Northumberland, wrote, "I will consider most carefully the documents you have sent me on the subject of the Druridge Bay Campaign, but I am most unlikely to be involved in the Bill in the House of Lords in any way whatever. I therefore do not think there is any purpose in discussing this with you at this time, and I have to say that I am, in principle, in favour of nuclear electricity for reasons which you will be well aware of. It is most unlikely I shall get involved in any way with the public debate that may happen about Druridge Bay, either for or against."

His letter demonstrated to me the difference between a representative in the Houses of Parliament who depends on being elected and one who does not. He was able to express his opinion quite frankly, without regard for local public opinion. A letter from an MP who did not support what he knew to be a sensitive local issue would have been differently expressed, equally courteous perhaps, but more ambivalent and guarded.

The date was fixed for 19 April 1989 for our presentation to be given at the House of Lords. This time, the various campaigners made separate travel arrangements rather than going in a coachload. Most of our group were people of modest means, used to pinching and scraping. Somehow we had to combine arriving at the House of Lords, looking suitably dignified, with our budget travel and accommodation arrangements. Fiona Hall, my daughter Laura and I went down the day before and stayed in a youth hostel near St Paul's Cathedral. Adrian Smith combined his journey with business for Newcastle City Council.

Gordon Steel, our treasurer, had various adventures before he arrived at the red carpets of the House of Lords. He had travelled overnight by bus from Newcastle. In his best clothes, rather crumpled from the journey, he was sitting in Hyde Park at 6.30 am on the day of the lobby, ready for his breakfast. He recalled, "I was really hungry. I had a tin of sardines which I was looking forward to eating but there were no instructions on how to open it. I started hacking at it with my penknife, but it didn't work. I got so frustrated I put it on the ground and jumped on it. It opened all right. The sardines and oil shot all over the place, including on my trousers. I thought, 'That's a good start to my visit to the House of Lords, a nice smell of sardines on my trousers.' So I jumped around in the Serpentine to wash my trousers as well as I could, and then ran around to get dried off."

Outside the House of Lords, April 1989, Adrian Smith is interviewed for the Druridge Bay Campaign by BBC Look North. *Photo Gordon Hanson.*

Later in the day, Gordon met his MP Ted Garrett in the House of Commons. Ted Garrett was the only North East Labour MP who was sitting on the fence on the Druridge issue, as he was sponsored by the Amalgamated Engineering Union which favours nuclear power. Gordon has had many arguments with him in the past, and brought the subject up again over more than a couple of drinks in the House of Commons bar, before they finally arrived in a very relaxed mood at the House of Lords committee room where we were meeting.

The same morning, before the lobby, Laura and I thought we would do some sightseeing so we walked miles through St James' Park and down Whitehall, ending up at Westminter Abbey. When we arrived Laura needed the toilet. Knowing that the nearest was a few blocks away, and tired of carrying our heavy rucksacks, we left them on some chairs. When we returned fifteen minutes later, there was a crowd outside the abbey, and police everywhere. We asked the people standing around what was the problem, and were told that two bags had been left unattended. The police were waiting for sniffer dogs. It was very embarrassing having to go out in front of all those people to the black robed cleric and tell him the bags belonged to us. Crowds of tourists followed us as we were escorted by the cleric and the police into the abbey.

Immediately after that incident, we had to appear cool, calm and collected to meet Lord Glenamara and the media. Lord Glenamara came out with Lord Williams of Elvel, opposition spokesman for energy in the House of Lords. Both they and Adrian were interviewed for north-east television viewers on Tyne Tees and BBC.

Eager family and friends back home watched the six o'clock news, Fiona's daughters among them. She was among the group in the background, while the VIPs were being interviewed. As soon as she reached home next day, her daughter Catherine said, "I saw you on television, Mummy. You were blowing your nose!"

Once inside the House of Lords, we were pleased to be joined by Lord Dormand (formerly Jack Dormand MP of Easington), Lord Irving, Lord Ross, and Lord Winstanley. Together with Lord Glenamara and Lord Williams, they listened carefully to our case and proposed amendments which we had tailored anew for the House of Lords.[7]

The lords present were very sympathetic, for various reasons. Lord Dormand, as a former local MP with interest in the coal industry, had long been a patron of the DBC. Lord Winstanley was a former chairman of the Countryside Commission, and had visited Druridge Bay in the past. Lord Irving of Dartford and Lord Ross of Newport had a strong interest in energy matters, the latter being one of the SLD spokesmen in the House.

Inside the House of Lords committee room, after giving their presentation, are, from left to right: Billy Smith, Ray Lashley and Dennis Murphy NUM, Lord Ross, Fiona Hall, Lord Williams of Elvel, Jane Gifford, Lord Dormand of Easington, Laura Gubbins age 9, Lord Glenamara (formerly Ted Short MP Newcastle Central), Lord Wistanley, Bridget Gubbins, Adrian Smith and Gordon Steel. *Photo Gordon Hanson.*

Lord Williams agreed to adopt our amendments in full. They would be discussed in the House of Lords in May and June. We kept in regular correspondence with the lords during the next couple of months.

The second reading debate in the House of Lords was on 25 April. We obtained Hansard, and contacted all the lords who spoke out for energy efficiency. By doing that, we were able to add Lord Hatch to our list of VIP supporters.

The amendments came up for discussion on 16 May. There was tension and excitement among campaigners, nationally and in the north east. Friends of the Earth in London had also been promoting an amendment to the bill's inadequate clause on energy efficiency, gaining their own contacts and support in the House of Lords. On May 17, the House of Lords embarrassed the government by voting with a majority of 12 for this energy efficiency amendment.[8]

I was in close contact with Energy Campaigner of Friends of the Earth, Simon Roberts, during this event, and he was thrilled with its success, which made national news.

One of our amendments on energy efficiency was discussed the next day, as were the others later. They were not voted on as the earlier FoE prompted one had been. It was disappointing but not devastating as we had achieved much in the attempt. We had contributed to the debate, enlisted more support, and Friends of the Earth's amendment had very similar aims to our own. We shared their pleasure in their success.

Further problems for Cecil Parkinson, Secretary of State for Energy, emerged as 45 back-bench MPs threatened a revolt in support of the Lords' amendment, and although the revolt was subdued, it had pointed to the weakness of the government's case. We drew attention to it locally by writing to our own Conservative MPs Alan Amos and Neville Trotter asking them to support the revolt and inform the newspapers.

Ultimately, the government's compromise amendment stated that the electricity industry regulator may "determine such standards of performance in connection with the promotion of the efficient use of electricity by consumers, as in his opinion, ought to be achieved by such suppliers". In practice this amounted to little more than making available leaflets about energy efficiency in high street showrooms. A major opportunity to reduce the number of new power stations and reduce pollution had been lost.

More embarrassment to the government occurred when the Energy Select Committee's report on the Greenhouse Effect was issued on 17 July. Sir Ian Lloyd, true blue Tory, harshly criticised the government's record on energy efficiency, and by implication the current electricity privatisation legislation. The 70 page report rejected Mr Parkinson's claim that the government should impose no changes to its energy policy and that global warming should be left to "market forces". It said that Britain needed a moral lead in fighting the

Greenhouse Effect, in order to enjoy the economic lead that new types of technologies would bring.

The importance of energy efficiency was thus emphasised by sturdy supporters of the present government, as well as by its critics. Its role in helping to control the Greenhouse Effect was clear. Although the Lords' energy efficiency amendment had been watered down by the House of Commons, the intellectual argument was being won.

We had more publicity for our cause, when on 14 July, in the House of Lords debate on the Electricity Bill, two lords specifically spoke up for Druridge Bay. Lord Williams was seeking to ensure that future privatised electricity companies shall have a duty to "further the conservation and enhancement of natural beauty." He concentrated his arguments on land in Snowdonia and Druridge Bay.

Then Lord Dormand spoke about Druridge Bay, pointing out how opencast sites there had been restored to farming and turned into nature reserves. He also said that after the 30-40 years of a nuclear power station's working life, what remains will be sealed and covered for about 100 years. He used a quote from one of the Druridge Bay Campaign's leaflets, "It sounds like a fairy tale, but that is the official CEGB policy."

As ever, we sent this information to local papers. People could see that the Druridge Bay issue was drawing national attention.

The final drama in the passage of the electricity bill occurred on 24 July, on the last day when Parliament was considering it, and when many of us were on holiday. Cecil Parkinson announced to an astonished House of Commons that the older Magnox nuclear power stations were not to be included in the electricity sell-off, but would remain in state hands. This was a complete reversal of the government's position, which until this point had been adamant that nuclear power was to be privatised along with the rest of the electricity industry. This was a visible crack in the wall of the nuclear industry.[9]

Earlier in July, a devastating report by city brokers UBS Philips and Drew had claimed that the unknown cost of decommissioning and dealing with nuclear waste could be more than the combined current value of the nuclear generating plant owned by the CEGB. In other words, the nuclear industry had a negative value. The government pledge of £2.5 billion to help cover these costs would be inadequate, and National Power could face a decommissioning bill of £12 billion.[10]

There is absolutely no doubt that this type of report had caused the government's radical shift. Nuclear power was not marketable. Despite Cecil Parkinson's determination to include it in the privatisation package, and despite his schemes to protect it from the full effect of market forces, the costs were going to be so high that it was clear nobody would invest in it. Thus he had been obliged to retain state ownership of the Magnox power stations. That was the thought we all took with us on our summer holidays.

By the time we produced our August newsletter, the bill had received Royal Assent, and was therefore law.

None of our amendments, or those we supported, became a part of the new Electricity Act, but our actions served other purposes. We had correctly isolated and campaigned on key issues including the protection of nuclear power, our right for a full public inquiry and the need for effective energy efficiency legislation. The media attention we received in the north-east kept our members, our supporting councils and the public well-informed about our aims, and those of us who had taken part in the exercise learned how to make a point of view heard politically.

• • •

After the 1989 summer break, Druridge Bay Campaign returned to its normal activities. In September, we held our annual fair. The deputy mayor and mayoress of Morpeth, Councillor and Mrs Clive Temple, attended. We also had a glossy new sign erected at The Cairn, to replace the shabby old one. Alan Beith unveiled it, saying, "The campaign to save Druridge Bay is now entering a critical phase because of the privatisation of the electricity industry. We might have won the first stage of the argument but forces in the electricity industry are still there which want nuclear power stations."

That autumn, news items appeared every few days with prominent people speaking out strongly for or against nuclear power. First the anti-nuclear lobby would gain a point, then the opposite.

Bad news for the nuclear industry had been the exposure in July of a leukaemia cluster near Hinkley, in a study carried out by consultant haematologists and Somerset Health Authority.[11] Then in August, the US government announced it would have to spend more than £10 billion over the next five years cleaning up its nuclear weapons plants to bring them into line with more stringent environmental controls.[12] Another blow came when it was announced that Wackersdorf, Germany's prospective Sellafield, was now to be used to produce solar energy cells and recycle BMW car wrecks, a triumphant success for environmental groups.[13]

Tony Blair in August revealed that the AGRs (the UK's second generation nuclear power stations) were performing even more badly than the Magnox reactors which Cecil Parkinson had removed from privatisation, and presented an even worse financial risk. Costs were likely well to exceed the £4.5 billion for decommissioning estimated for the Magnox reactors.[14]

At the Hinkley Inquiry, the cost of building Sizewell B was revealed to be even higher, 10% more than estimated. This was a further blow to the new Secretary of State for Energy, John Wakeham, who was carrying out high-level discussions on the contracts for the new electricity industry.[15]

Despite these problems, the nuclear proponents were still on the offensive. In October, Sam Goddard of the fledgling National Power, the new private company which would be taking over the CEGB's nuclear component, was speaking at a Sheffield meeting of the British Association. He confirmed the argument that more nuclear power and energy conservation would be needed to help protect the environment from the Greenhouse Effect. Improvements to new and existing coal fired power stations would not be enough. If Britain was to cut carbon dioxide emissions from power stations to the level achieved by France, there would have to be sevenfold increase in the nuclear programme.[16]

Similarly, in October, John Wakeham was still re-inforcing the government's traditional line on nuclear power and privatisation. At the Conservative Party Conference, he said by phasing out nuclear power Labour would also destroy thousands of jobs. "I for one will never put those jobs at risk, or gamble with Britain's future. While I am Secretary of State for energy, I will not allow Britain to be held to ransom again ... by the militants in the NUM."[17]

Then Lord Marshall, pro-nuclear chairman of the CEGB, admitted that the true figure for decommissioning of Magnox reactors was £6.6 billion, rather than the previous CEGB estimate of £2.8 billion. This expense would be paid by the taxpayer as these reactors were not to be sold after all, while the profitable part of the electricity industry would be sold to shareholders. Tony Blair described Lord Marshall's speech as "the most extraordinary indictment of the government, the nuclear industry and this chaotic privatisation."[18]

On 26 October, *Guardian* reporter David Fairhall commented on the changing climate and attitudes to nuclear power. He said, "Astonishing as it may seem the woolly-pullovered environmentalist's staunchest ally in the nuclear debate is turning out to be the sharp-suited city accountant. While the pro-nuclear lobby has undoubtedly found a powerful argument in the Greenhouse Effect, it will not suppress public unease until it can finally resolve the two fundamental issues of safety and waste disposal. The accountants considering the prospects for commercial nuclear power are terrified by what they see."

The most serious indication so far of what was happening to undermine nuclear power came in the *Observer* headlines of 29 October, "End of the road for nuclear power in the UK". It said, "Plans to build a series of pressurised water reactors are under imminent threat of cancellation... This blow to the country's nuclear industry is the direct consequence of soaring estimates in the price of nuclear generated electricity... In particular decommissioning costs have risen from the £3 billion estimated by former Energy Secretary Cecil Parkinson last year to a current figure of £15 billion, a sum equivalent to the price-tag of the entire UK electricity industry," and "Any remaining hope of privatising the nuclear power stations of Britain must by now be entirely forlorn".

In the same article, it was suggested that plans for further reactors after Sizewell B, Hinkley Point, Wylfa in Anglesey and Sizewell C would be scrapped.

Nevertheless National Power was still optimistic of obtaining government support. On 7 November, at a European conference on the PWR, Alan Walker, engineering director, said that plans to bring four PWRs on stream by the turn of the century would depend on winning financial support from the government. With Sizewell B under construction, National Power would push ahead to build three more PWRs of the same design. He anticipated a favourable outcome of the Hinkley Point inquiry, and said that the "first structural concrete" could be laid by summer 1991.[19]

Guardian industrial correspondent Patrick Donovan wrote on 7 October that "his pledge that four of the latest generation reactors would be on-stream by the turn of the century seemed to owe more to wishful thinking that any realistic assessment of the huge problems confronting the UK nuclear industry at this crucial stage of its development".

Despite all these criticisms, we had no illusions about the government's commitment to nuclear power. Mrs Thatcher had become an "environmentalist" when finally convinced of the threat of global warming. Her conversion to "green" views on energy was convenient because she could argue against coal which she thoroughly disliked, and in favour of nuclear power of which she was an ardent advocate.

On 8 November, Mrs Thatcher made a speech on the global environment at the UN General Assembly. She said, "nuclear power – despite the attitude of so-called greens – is the most environmentally safe form of energy." Talking about UK programmes to reduce pollution, she said, "we shall be looking more closely at the role of non-fossil fuel sources, including nuclear, in generating energy."

• • •

9 November 1989 was a day to remember. The Berlin Wall came down. A monument to centralised power, which for decades had seemed permanent and invincible, had cracked, split open, and unarmed people had come pouring through. At breakfast time, I looked over the front page of the *Guardian*, which was dominated by photos and stories from Germany.

At the bottom of the page was the headline, "N-plant sale set to be scrapped". I was amazed, stunned. Immediately I phoned Jane Gifford. We communicated our shock and surprise to each other. The nuclear industry was not to be privatised, which was a humiliating come-down for the government. More astonishingly, the report also said, "Ministers are expected to scrap plans for four pressurised water reactors on the grounds that the latest technology is no longer economically viable."

No more nuclear power stations would be built! They were not economic! No-one would buy existing ones, therefore the state must keep them!

Another wall had collapsed, that of the massive nuclear power industry which for too long had held its secrets to itself. What had been gradually revealed over the last year, in the course of privatisation of electricity, had been too shocking even for Mrs Thatcher's pro-nuclear government to accept. Private profits and the nuclear industry were incompatible. Two of the central tenets of the Conservative government had clashed and caused a collapse.

I went to the DBC office and rang Alan Beith and Jack Thompson at the House of Commons. They were both to be attending the speech by John Wakeham at 4 pm that day and each promised to keep in close contact. I rang the local TV and radio news desks to alert them to the significance of the news for Druridge Bay. They did not respond at first as I expected. The Berlin Wall story was dominating everything.

In the afternoon I was digging in the garden. Years of concern and campaigning seemed to be reaching a crucial point and it was hard to keep a steady head. At 3 pm the phone rang. It was Alan Beith. He had an advance copy of John Wakeham's speech.[20] He said, "We've won, Bridget! I wanted you to be the first to know!" Shortly after that he was interviewed at Westminster by BBC's Look North, as was Ronnie Campbell MP for Blyth Valley.

At 4 pm, when I was still gardening, Tyne Tees Television telephoned, asking if I would be prepared to talk on the six o'clock news. I was happy to agree, and rushed to scrub the mud from my fingers. Then Jack Thompson rang, counselling caution. Even though it looked like there would be no more nuclear power stations for some time, he warned that the nuclear industry was still a powerful and dedicated lobby. He had heard the speech at 3.55 pm, and reminded me that the government had promised a review of nuclear power in 1994. Tyne Tees sent a taxi for me, and I did a studio interview, participating in the live production of Northern Life.

I had the words of both MPs ringing in my ears as I spoke, together with their warning that Druridge Bay would not be safe until the land was out of the hands of the nuclear industry.

• • •

9 November 1989 brought to an end the first epoch of campaigning to save Druridge Bay from the threat of nuclear power stations. The downfall of the civil nuclear power industry had been caused by a combination of its own innate problems and their exposure by its critics. The CEGB's monopoly of power, control of information and selective accounting procedures had hidden its environmental costs for a decade. These were exposed by city accountants during the privatisation process, and their criticisms carried more weight with the government and shareholders than arguments by anti-nuclear groups. Fears

of Chernobyl-type accidents or the need to store radioactive waste for hundreds of thousands of years had not been as ultimately damaging as the fact that nuclear power could not make a short term profit for shareholders. In the Thatcher years, through which the Druridge Bay Campaign's work had so far taken place, nuclear power's non-profitability had caused its downfall.

The Druridge Bay Campaign had added its voice to those arguing against nuclear power nationally and globally. Focussed on the need to defend our own corner of the world, we had learned about the full potential impact of nuclear power, had our horizons widened and acted upon what we knew. Ours had been a modest contribution, but one which had a strongly rooted sense of purpose.

For over a decade, we had built up our pressure group and generated the energy to protect the bay. But our work is not over. The pressure must be maintained at an even and controllable level for many years to come. The nuclear industry is not lying down and dying. Nuclear Electric has refused to sell the land at Druridge Bay. The state-owned company, though currently maintaining a low profile, is quietly, stubbornly active, and is patient and resourceful. The Druridge Bay Campaign must be the same.

NOTES

1. The wording of the three lobbying points was as follows.

THE BILL SHOULD NOT GIVE NUCLEAR POWER A PROTECTED STATUS. This would lead to more nuclear power stations, including Druridge Bay, displacing better alternatives.

FUTURE PUBLIC INQUIRIES INTO NUCLEAR PROPOSALS WOULD BE SIGNIFICANTLY RESTRICTED IN SCOPE, as a result of the protected status of nuclear power. It would limit the opportunity of North East people to defend their area.

ENERGY EFFICIENCY IN GENERATION AND USE OF ELECTRICITY SHOULD BE A STATUTORY REQUIREMENT PLACED UPON THE NEWLY PRIVATISED INDUSTRY. Reduced demand would mean fewer controversial new nuclear power stations.

i. **The protected status of nuclear power.** A certain percentage of electricity generated (assumed to be about 20%) was to be obtained from nuclear power. This requirement was known as the Non-Fossil Fuel Obligation or NFFO.

The figure of 20% was based on the amount of electricity which nuclear power was likely to supply. In 1988, nuclear electricity was providing about 14% of UK electricity. The difference was based on the expected contribution from new nuclear power stations, plus a small amount from renewable sources.

This nuclear generated electricity would be more expensive than that from coal or gas fuelled plants. Customers would thus be paying a supplement to support the nuclear industry, and in our case perhaps be subsidising a future unwanted nuclear power station at Druridge Bay.

The protection of nuclear power would also mean a slowing down of building new, cost-effective, cleaner power stations. Adrian Smith was involved in Newcastle City Council's efforts to build a combined heat and power station.

ii. **The public enquiry** Northumberland County Council and the Druridge Bay Campaign had been concerned that any future public inquiry into nuclear power at Druridge Bay should be wide ranging. It should include the effect on the coal industry, and local need for electricity. It should not be limited to local planning issues, such as the colour of the power station, tree planting etc. Northumberland County Council spent £60 000 at the Sizewell Inquiry, and believed it had elicited government assurance of a wide-ranging planning inquiry at Druridge Bay. Adrian Smith's reasoning was that if the Privatisation Act should be passed, the *need* for nuclear electricity would be enshrined by law via the NFFO. Also fossil-fuelled power stations could not be argued as an alternative to nuclear. Economic decisions would be taken by private companies, and would not enter into the scope of a public inquiry.

iii. **Energy Efficiency** The Electricity Act presented a great opportunity to embody the saving of electricity in legislation. Many utilities in the USA were regulated in such a way that it was more profitable for them to encourage customers to reduce electricity consumption than to build more new capital-expensive power stations to satisfy an endlessly increasing demand. But the British Government seemed satisfied with an electricity industry which would sell as much electricity as possible in order to make the greatest profits. This totally clashed with newly emerging ideas on slowing down global warming by reducing inefficient use of fuel and electricity, and excess consumption.

2. At the Hinkley Inquiry, Frank Jenkin of the CEGB said that at the 5% rate of return expected by public sector schemes, nuclear was cheaper than coal, at 2.4 pence per kilowatt hour compared to 2.5. However, if the rate of return was fixed at the higher level expected in the private sector, between 8% and 10%, nuclear's advantage disappeared. At 10% he said, "the economics of a new nuclear station are significantly adverse against all the alternatives." *Guardian* 7.12.88

Dr Swift-Hook for Stop Hinkley Expansion told the inquiry that wind turbines could supply the equivalent of half the capacity of Hinkley C at half the price of the £1.5 billion programme. The unit cost of wind generated electricity at 2.2 pence per kilowatt hour would also be cheaper than nuclear power. *Guardian* 14 11 88

3. *Guardian*, 1 4 89

4. *Guardian*, 30 3 89

5. The CEGB gave evidence to the House of Commons Energy Select Committee in its examination of the energy policy implications of the Greenhouse Effect. They calculated that one PWR (the size and type proposed for Druridge Bay) would abate anticipated carbon dioxide emissions from an equivalent coal fired power station by six million tonnes per year. They expected that by the year 2000, the nuclear industry would have demonstrated its capacity to build one PWR per year. Thereafter, the rate could be increased to two PWRs per year, or possibly more. They calculated that by the year 2005, a further 10 PWRs could be commissioned. (*ENERGY POLICY IMPLICATIONS OF THE GREENHOUSE EFFECT*, HMSO, P 27, 8 2 89) One of these would certainly be at Druridge Bay, as it is one of the few suitable sites in the country.

6. New evidence to the Hinkley Inquiry by Dr Nigel Mortimer for Friends of the Earth showed that nuclear power caused the emission of a good deal of carbon dioxide, not so much in generation as in extraction and processing of ore. With a world-wide replacement of coal-fired power stations by nuclear, high grade uranium would quickly run out. Low grade uranium requires large amounts of fossil fuel to mine and process it. Once uranium ore grades fall to between 1 and 100 parts per million uranium oxide, carbon dioxide levels from coal and nuclear power station cycles would be the same. If nuclear power were used as a major source of electricity worldwide, supplies would run out in as little as 23 years.

7. The amendments were to promote the following points:
- that a new large power station should not be built if it is more expensive to do so than investing in saving electricity.
- that reductions in demand for electricity brought about by energy efficiency should be included in the non-fossil fuel obligation
- that planning inquiries into new power stations should be held in public, and must include debate on whether or not the power station is needed, despite the ruling that a percentage of electricity must come from nuclear power.

8. Lord Shepherd proposed the amendment, which was supported by Conservative peers, the Earl of Lauderdale and Lord Peyton. The amendment obliged companies to promote efficient use of electricity. If they did not, the Energy Secretary could refuse to allow them to put up electricity charges or apply for major capital projects. Lord Shepherd argued that if proper conservation measures were in force, four out of five new power stations, costing £10 billion, would be postponed.

9. Cecil Parkinson's reasoning was as follows. The Magnox stations were drawing to the end of their lives. One was already closed, and most of the others were due to close within the next few years. "Most of these costs therefore relate to the past, to electricity already generated and paid for," he said. "Future customers will be bearing the full cost of the electricity they consume. It would not be right to burden them also with costs arising from the past.

"National Power will also be building new pressurised water reactors. In order to enable the nuclear generating companies to focus their attention on the future, the government has decided that it would be appropriate to relieve the new companies of dealing with these substantial problems of the past."

The government planned to increase from £1 billion to £2.5 billion the fund which was to be allocated to "unforeseen circumstances", he said. This fund was intended to ease the problems caused in the future by unknown costs of disposal of radioactive waste and decommissioning of the nuclear reactors which were still to be privatised, the AGRs and the new PWRs.

Tony Blair was astonished by Cecil Parkinson's announcement. He said, "The Magnox stations remain in public ownership, so the full costs of decommissioning, disposing and reprocessing of the waste will fall on the taxpayers. The latest estimates are £500 million a station. A total of £4.5 billion, perhaps even more." *Guardian* 25 7 89

After Cecil Parkinson's statement, Tony Blair sought postponement of the bill. He said months of consideration of the bill in standing committee were now useless. His motion was lost by a government majority of 107.

10. *Guardian*, 18 7 89
11. *Guardian*, 28 7 89
12. *Guardian*, 1 8 89
13. *Guardian*, 8 8 89
14. *Guardian*, 9 8 89
15. *Guardian*, 26 9 89
16. *Guardian*, 15 9 89
17. *Guardian*, 11 10 89
18. *Guardian*, 18 10 89

19. *Guardian,* 7 11 89

20. On 9 November 1989, John Wakeham, Secretary of State for Energy, made the following statement to the House of Commons at 3.55 pm. "The government have concluded that the English advanced gas cooled reactors and Sizewell B should remain, along with the Magnox stations, in a government-owned company." In other words, the nuclear power stations would not be privatised. His reason was that, during discussions about the financing of new nuclear power stations, unprecedented financial guarantees were being sought of the government. He said, "I am not willing to underwrite the private sector in this way".

He also said, "The Non-Fossil Fuel Obligation will be set at a level which can be satisfied without the construction of new nuclear stations beyond Sizewell B." This meant in practice that the Non-Fossil Fuel Obligation (NFFO) would be set at the level which existing nuclear power stations plus Sizewell B could supply. New nuclear power stations would not need to be built to satisfy the NFFO.

He made it clear that nuclear power was still favoured by the government for reasons of diversity of supply, and that "by maintaining the nuclear option we are creating the opportunity for a longer-term contribution from economic nuclear power".

"I am asking the CEGB to consider urgently what action it wishes to take with respect to its applications for my consent to build PWR stations at Hinkley Point C, Wylfa B and Sizewell C". This effectively meant that there would be no government financial backing for these proposals at present.

He said, "The government will wish to review the prospects for nuclear power as the Sizewell B project nears completion in 1994".

POSTSCRIPT

Following the announcement by John Wakeham on 9 November 1989, the Druridge Bay Campaign had to make an important decision. Should the organisation be almost folded up, closing the office, and most members relaxing their commitment? Should it go on, business as usual, concentrating on the issue of nuclear power at Druridge Bay, even though the immediate threat had been removed? Or should it retain continuous vigilance while developing in other directions?

Much internal debate took place over the following three months. In March 1990, at an Extraordinary General Meeting of the DBC at County Hall, Morpeth, members and affiliated groups voted overwhelmingly to continue with a vigorous campaign to protect Druridge Bay from future threat of nuclear power stations. They also decided to change the DBC's emphasis to fit the new circumstances and the time lag in which there will probably be no planning application for nuclear power stations.

Influential in this decision had been the following letter to Councillor Kevin Flaherty, Chairman of Planning and Economic Development at Northumberland County Council, from Sam Goddard of Nuclear Electric, dated Feb 5 1990.

"Nuclear sites are difficult to find and Druridge, because it offers an acceptable balance of technical and environmental factors, is considered by Nuclear Electric to be a valuable asset which it will be retaining for possible future nuclear power station development. When in the future the case for further nuclear power stations is established, Druridge will be among the site options available for consideration by Nuclear Electric."

Similar letters were sent to Alan Beith and Jack Thompson. Nothing could be clearer than that Druridge Bay was still under threat, albeit not in the immediate future. I therefore strongly believe that the DBC made the right decision in its move to maintain vigilance.

It was also agreed that the DBC's remit should be widened. While retaining its primary reason for existence as opposition to nuclear power proposals, its new remit became as follows:

"Druridge Bay Campaign works to oppose nuclear power at Druridge Bay by enhancing the beauty of the bay and promoting clean, safe electricity supply for the region without nuclear power."

There were two new series of actions made possible by the widening of the remit. Firstly, other environmental threats to Druridge Bay could be tackled. The burning issue here was the extraction of sand from the dunes and beach by Ready Mixed Concrete's subsidiary Northern Aggregates. Previously, despite the concern of many of our members, our remit did not extend to this problem.

Secondly, positive campaigning for safe, non-nuclear energy policy would no longer be inhibited by the rigid adherence to "opposition to nuclear power at Druridge Bay". While this remained the central tenet of our existence, we were free to range wider in order to achieve our goal and promote a desirable energy policy for the North East. If electricity demand continued to increase at the then current rate of 1.9% per annum, a new power station equivalent to the size of that proposed for Druridge Bay would be needed by the year 2001. When Blyth coal-fired power station closes early next century, supply equivalent to a second large nuclear power station will be needed. The DBC sees it as essential that a non-nuclear regional energy policy is developed to prevent nuclear power stations being built at Druridge Bay.

This aim must be achieved by a combination of reduced energy need and new non-nuclear supply. The former requires improvements in conservation and energy efficiency. The latter requires new developments in combined heat and power, wind, wave, small scale hydro, biomass, waste burning and geothermal systems.

The wider remit also permitted us to take part in the national environment/energy debate, which could include expressing views on nuclear power.

Two working groups within the DBC are now campaigning on the issues of both sand extraction and energy policy. Like the Radiation Monitoring Group, they are specialist sub-groups working under the overall direction of the elected Steering Committee.

Meanwhile, apart from a series of advertisements in the national press continuing to promote the virtues of nuclear power, Nuclear Electric has gone very quiet. Little is said about the 1994 Review of nuclear power promised by John Wakeham.

Ominous signs for the future can be glimpsed in plans being developed by British Nuclear Fuels. Sites for new PWRs are being investigated in detail at Calder Hall near Sellafield and Chapel Cross, near Dumfries. BNFL is acting in conjunction with Nuclear Electric, Siemens of Germany and Framatome of France. What is to stop Nuclear Electric making a deal with these or other companies to develop Druridge Bay?

East Anglia Daily Times on 26 February 1991 reported that Nuclear Electric is arguing that there is a strong economic case for re-starting the construction programme as soon as possible. It wants to proceed with Sizewell B's first twin at Hinkley Point in Somerset, and with Sizewell C. A fourth PWR would follow at Wylfa in Anglesey. The paper reports that plans for Sizewell C are at such an advanced stage that construction could start at any time, and Nuclear Electric wants to take advantage of the expertise and machinery now assembled in Suffolk for Sizewell B construction.

In a letter to the DBC on 17 April 1991, Nuclear Electric wrote, "We continue to regard Druridge Bay as a suitable candidate site for further nuclear developments. The technical criteria required for nuclear sites are very onerous and locations such as Druridge, which offer an acceptable balance of technical and environmental factors, are difficult to find".

Children growing up in the 1980s have had the shadow of nuclear power at Druridge Bay hovering over them. It may be that they themselves will be tackling the nuclear industry as they grow into adults towards the end of the 90s. Perhaps there will be another story like this in a decade's time, with a conclusive ending one way or another. Their vigilance as well as ours will be required.

The Cairn 1987. *Photo Peter Berry*